NELSON VICscience

Andrea Blunden

Leigh Park

SKILLS WORKBOOK
psychology VCE UNITS ③ + ④ 4E

VICscience Psychology VCE Units 3 & 4 Workbook
4th Edition
Andrea Blunden
Leigh Park
ISBN 9780170465069

Publisher: Eleanor Gregory
Content developer: Katherine Roan
Project editors: Felicity Clissold, Alex Chambers, Alana Faigen
Copyeditor: Lauren McGregor
Proofreader: Vanessa Lanaway
Series text design: Ruth Comey (Flint Design)
Series cover design: Emilie Pfitzner, Everyday Ambitions
Series designer: Cengage Creative Studio
Permissions researcher: Helen Mammides, Alex Chambers
Production controllers: Karen Young, Alex Chambers
Typeset by: Lumina Datamatics

Any URLs contained in this publication were checked for currency during the production process. Note, however, that the publisher cannot vouch for the ongoing currency of URLs.

Acknowledgements
Extracts from the VCE Psychology Study Design (2023-2027) are used by permission, © VCAA. VCE® is a registered trademark of the VCAA. The VCAA does not endorse or make any warranties regarding this study resource. Current VCE Study Designs, past VCE exams and related content can be accessed directly at www.vcaa.vic.edu.au.

We would like to acknowledge the following for permission to reproduce copyright material - images for each chapter opener:
Ch 1: Shutterstock.com/Monkey Business Images
Ch 2: Shutterstock.com/FamVeld
Ch 3: Adobe Stock/WavebreakMediaMicro
Ch 4: Adobe Stock/Maria
Ch 5: Adobe Stock/ Irina Schmidt
Ch 6: Shutterstock.com/Monkey Business Images
Ch 7: Adobe Stock/peopleimages.com
Ch 8: Shutterstock.com/Jag_cz
Ch 9: Adobe Stock/Valerii Honcharuk
Ch 10: Shutterstock.com/fizkes
Ch 11: Adobe Stock/aFotostock

© 2023 Cengage Learning Australia Pty Limited

Copyright Notice
This Work is copyright. No part of this Work may be reproduced, stored in a retrieval system, or transmitted in any form or by any means without prior written permission of the Publisher. Except as permitted under the *Copyright Act 1968*, for example any fair dealing for the purposes of private study, research, criticism or review, subject to certain limitations. These limitations include: Restricting the copying to a maximum of one chapter or 10% of this book, whichever is greater; providing an appropriate notice and warning with the copies of the Work disseminated; taking all reasonable steps to limit access to these copies to people authorised to receive these copies; ensuring you hold the appropriate Licences issued by the
Copyright Agency Limited ("CAL"), supply a remuneration notice to CAL and pay any required fees. For details of CAL licences and remuneration notices please contact CAL at Level 11, 66 Goulburn Street, Sydney NSW 2000,
Tel: (02) 9394 7600, Fax: (02) 9394 7601
Email: info@copyright.com.au
Website: www.copyright.com.au

For product information and technology assistance,
in Australia call **1300 790 853**;
in New Zealand call **0800 449 725**

For permission to use material from this text or product, please email
aust.permissions@cengage.com

ISBN 978 0 17 046506 9

Cengage Learning Australia
Level 5, 80 Dorcas Street
Southbank VIC 3006 Australia

Cengage Learning New Zealand
Unit 4B Rosedale Office Park
331 Rosedale Road, Albany, North Shore 0632, NZ

For learning solutions, visit **cengage.com.au**

Printed in China by 1010 Printing International Limited.
1 2 3 4 5 6 7 26 25 24 23 22

Contents

Introduction vi
Key science skills grid vii

1 Scientific research methods — 2

1.1 The process of psychological research investigations 2
    1.1.1 Developing your research question, aim and hypothesis 2
    1.1.2 Predicting outcomes 4
    1.1.3 Sampling techniques 5

1.2 Scientific investigation methodologies 6
    1.2.1 Methodologies 6

1.3 The controlled experiment in detail 7
    1.3.1 Variables in controlled experiments 7
    1.3.2 Experimental designs for controlled experiments 9

1.4 Analysing and evaluating research 10
    1.4.1 Displaying data in tables, bar charts and line graphs 10
    1.4.2 Descriptive statistics 12
    1.4.3 Quality of data 13
    1.4.4 Know what you are reading! 15
    1.4.5 What is your data telling you? 16
    1.4.6 Ethics 18
    1.4.7 Know your key terms 19

Exam practice 20

2 Nervous system functioning — 23

2.1 The nervous system: roles and subdivisions 23
    2.1.1 Nervous system 'What am I?' 23
    2.1.2 Nervous system concept map 26
    2.1.3 Evaluation of research 29

2.2 Conscious and unconscious responses 31
    2.2.1 Conscious and unconscious responses of the nervous system 31

2.3 The transmission of neural information 33
    2.3.1 The structure and function of neurons 33
    2.3.2 Neurotransmitters 34
    2.3.3 Nervous system 'match the pairs' 35

2.4 Neural basis of learning and memory 37
    2.4.1 Neurological basis of memory and learning 37

Exam practice 39

3 Stress as an example of a psychobiological process — 42

3.1 Stress and stressors 42
    3.1.1 Introduction to stress 42
    3.1.2 Sources of stress 45
    3.1.3 Flight-or-fight-or-freeze response 47
    3.1.4 Effects of stress 49

3.2 Selye's General Adaptation Syndrome 51
    3.2.1 The General Adaptation Syndrome 51
    3.2.2 The GAS: a case study 53

3.3 Stress as a psychological process 55
    3.3.1 Lazarus and Folkman's Transactional Model of Stress and Coping 55

3.4 The gut–brain axis 57
    3.4.1 The gut–brain axis (GBA) 57
    3.4.2 The human microbiome project 59

3.5 Coping strategies 61
    3.5.1 Strategies for coping with stress 61
    3.5.2 Evaluation of research 63

Exam practice 66

4 Approaches to understand learning — 70

4.1 Approaches to understand learning 70
    4.1.1 Different approaches 70

4.2 Behaviourist approaches to learning 71
    4.2.1 Classical conditioning process 71
    4.2.2 Classical conditioning key terms 73
    4.2.3 Ethics in learning research: 'Little Albert' 76
    4.2.4 Operant conditioning 77
    4.2.5 The ABCs of operant conditioning 79
    4.2.6 Using operant conditioning in animal training 80
    4.2.7 The psychology of advertising 81

4.3 The social-cognitive approach to learning 82
    4.3.1 Observational learning 82
    4.3.2 Application of research methods 83

4.4 Aboriginal and Torres Strait Islander peoples' approaches to learning 86
  4.4.1 The situated multimodal systems approach to learning 86
Exam practice 88

5 The psychobiological process of memory — 92

5.1 What is memory? 92
  5.1.1 Atkinson and Shiffrin's multi-store model of memory 92
5.2 The structure of long-term memory 93
  5.2.1 Concept map of memory 93
  5.2.2 Comparing implicit and explicit memories 96
5.3 The neural basis of explicit and implicit memories 96
  5.3.1 Brain structures involved in long-term memory 96
5.4 Autobiographical memory 98
  5.4.1 Studying aphantasia 98
5.5 Mnemonics 100
  5.5.1 Mnemonics: acronyms and acrostics 100
  5.5.2 Mnemonics: method of loci 101
  5.5.3 Research into memory 102
Exam practice 104

6 The demand for sleep — 108

6.1 What is consciousness? 108
  6.1.1 Types of sleep 108
  6.1.2 Brainwaves 112
  6.1.3 Physiological changes and stages of sleep 114
6.2 Regulation of sleep–wake patterns 116
  6.2.1 Circadian rhythms 116
  6.2.2 Stages of sleep 118
  6.2.3 Evaluation of research 119
6.3 The changes in sleep over the life span 121
  6.3.1 Investigating sleep 121
  6.3.2 Review of sleep 124
  6.3.3 Case studies 126
Exam practice 128

7 Importance of sleep in mental wellbeing — 131

7.1 Partial sleep deprivation 131
  7.1.1 Sleep deprivation 131
  7.1.2 A comparison between the effects of sleep deprivation and alcohol consumption 132

7.2 Sleep disorders 133
  7.2.1 Circadian phase disorders and shifts in the adolescent sleep–wake cycle 134
  7.2.2 Sleep–wake shifts in adolescence 135
  7.2.3 The effects of shift work on the sleep–wake cycle 136
  7.2.4 Bright light therapy and how to sleep better 137
7.3 Improving sleep–wake patterns and mental wellbeing 140
  7.3.1 Improving sleep hygiene 140
  7.3.2 Analysis of a research investigation 141
  7.3.3 Zeitgebers 142
  7.3.4 The importance of sleep to mental wellbeing – key terms 143
Exam practice 145

8 Defining mental wellbeing — 147

8.1 Ways of considering mental wellbeing 147
  8.1.1 Understanding key terms 147
  8.1.2 Resilience 149
  8.1.3 Factors that contribute to positive social and emotional health and wellbeing 150
  8.1.4 Social and emotional wellbeing (SEWB) 151
8.2 Mental wellbeing as a continuum 152
  8.2.1 Mental wellbeing 152
  8.2.2 Biopsychosocial approach to mental wellbeing 154
  8.2.3 Stress, anxiety and phobia 155
  8.2.4 Analysis of a research investigation 157
  8.2.5 Defining mental wellbeing concepts 158
Exam practice 163

9 Application of a biopsychosocial approach to explain specific phobia — 165

9.1 Development of specific phobia 165
  9.1.1 Factors that contribute to the development of specific phobia 165
  9.1.2 The biopsychosocial framework: application to understanding specific phobia 166
  9.1.3 Development of specific phobia 168
  9.1.4 Behavioural factors influencing the development of specific phobia 170

9.2	**Evidence-based interventions 171**	
	9.2.1 Terms used in evidence-based interventions 171	
	9.2.2 Understanding evidence-based interventions 172	
	9.2.3 Systematic desensitisation 173	
	9.2.4 Biopsychosocial approach study cards 174	
	9.2.5 Analysis of a research investigation 181	
Exam practice **183**		

10 Maintenance of mental wellbeing — 186

10.1 The application of a biopsychosocial approach to maintaining mental wellbeing 186
  10.1.1 The application of a biopsychosocial approach to maintaining mental wellbeing 186
  10.1.2 Australian Guide to Healthy Eating 187
  10.1.3 Adequate sleep 188
  10.1.4 Analysis of a research investigation 189
  10.1.5 Community support organisations visual presentation 190

10.2 Cultural determinants of social and emotional wellbeing 191
  10.2.1 The maintenance of wellbeing in Aboriginal and Torres Strait Islander peoples 191
  10.2.2 Aboriginal and Torres Strait Islander peoples' cultural determinants of health 192
  10.2.3 Maintenance of mental wellbeing concepts study cards 192

Exam practice 197

11 Using scientific inquiry — 199

11.1 Designing an investigation 199
  11.1.1 Formulating hypotheses 199
  11.1.2 Extraneous variables of all shapes and sizes 201
  11.1.3 Which experimental design is which? 203
  11.1.4 Sampling procedures 204
  11.1.5 Ethics in research 206

11.2 Conducting an investigation 206
  11.2.1 Experimental data 207
  11.2.2 Applying your knowledge 210

11.3 Science communication 211
  11.3.1 Analysing research 212
  11.3.2 Scientific terminology 215

Exam practice 239

Introduction

Psychology is a multifaceted discipline that seeks to describe, explain, understand and predict human behaviour and mental processes (VCAA VCE *Psychology Study Design 2023–2027*, page 6). Like the study of any other science, there is key knowledge and terminology that you need to know and understand and be able to use appropriately. However, no study of Psychology would be complete without also addressing the key science skills. The study of Psychology is not just about learning content, it is also about developing, using and demonstrating the skills that enable you to fully understand, experience and engage with the subject. It is about learning to think and work like a scientist.

Seven key science skills have been mandated by the Victorian Curriculum Assessment Authority (VCAA) across all VCE science subjects. These skills are transferable across subjects as well as being examinable in the VCE exam. Developing these key science skills means that you will be able to:

- develop aims and questions, formulate hypotheses and make predictions
- plan and conduct investigations
- comply with safety and ethical guidelines
- generate, collate and record data
- analyse and evaluate data and investigation methods
- construct evidence-based arguments and draw conclusions
- analyse, evaluate and communicate scientific ideas.

(VCAA VCE *Psychology Study Design 2023–2027*, pages 12–13)

Each of the key science skills listed above is broken up into multiple sub-skills. The mapping provided on pages vii–ix of this workbook allows you to see how these skills and sub-skills have been addressed in this workbook.

This workbook is full of activities that have been carefully crafted to enable you to consolidate your knowledge on a topic and to develop, use and demonstrate key science skills. Developing any skill takes time and practice; the key science skills in this book have been introduced in a graduated way starting with **practising** skills that you will have met in previous years of science study. As you gain proficiency and confidence, you will go on to **reinforce** newer and more complex skills. Then there are the new skills requiring an increased level of proficiency and thinking that you will **develop** during the course.

 Practise — Shows you activities that require previously introduced skills and will require practice as you work through the activities.

 **Reinforce** — Shows activities that will build on previously introduced skills.

 **Develop** — Shows activities that introduce a new skill or skills that require development and challenge you at a high level of proficiency.

This workbook can be used with any VCE Psychology resource that covers the VCAA VCE *Psychology Study Design 2023–2027*. It has been mapped to *VICscience Psychology Units 3 and 4* using icons in the student textbook. The icons have been placed in the textbook notifying you of the activities and the best place to undertake each activity.

The major headings in the student textbook match the major headings in *VICscience Psychology Units 3 and 4*. Applicable key knowledge is listed under each of the major headings. Skill activities have the applicable key science skills listed that students will be using or demonstrating.

Enjoy your study of VCE Psychology and take the time to develop, use and demonstrate the key science skills that are an integral part of this course.

Key science skills grid

Key science skill		1	2	3	4	5	6	7	8	9	10	11
Develop aims and questions, formulate hypotheses and make predictions	identify, research and construct aims and questions for investigation	1.1.1	2.1.3	3.5.2	4.3.2	5.5.3		7.3.2	8.2.4	9.2.5	10.1.4	
	identify independent, dependent, and controlled variables in controlled experiments	1.3.1	2.1.3	3.5.2	4.3.2	5.5.3	6.2.3	7.2.4 7.3.2	8.2.4		10.1.4	11.1.1 11.2.2
	formulate hypotheses to focus investigation	1.1.1	2.1.3	3.5.2		5.5.3	6.2.3 6.3.1	7.2.4 7.3.2	8.2.4		10.1.4	11.1.1 11.2.2
	predict possible outcomes of investigations	1.1.2										
Plan and conduct investigations	determine appropriate investigation methodology: case study; classification and identification; controlled experiment (within subjects, between subjects, mixed design); correlational study; fieldwork; literature review; modelling; product, process or system development; simulation	1.2.1 1.3.2	2.1.3	3.5.2	4.3.2							11.1.3
	design and conduct investigations; select and use methods appropriate to the investigation, including consideration of sampling technique (random and stratified) and size to achieve representativeness, and consideration of equipment and procedures, taking into account potential sources of error and uncertainty; determine the type and amount of qualitative and/or quantitative data to be generated or collated	1.1.3	2.1.3	3.5.2			6.3.1					11.1.4 11.2.1
Comply with safety and ethical guidelines	demonstrate ethical conduct and apply ethical guidelines when undertaking and reporting investigations	1.4.6	2.1.3	3.5.2	4.2.3 4.3.2							11.1.5
	apply relevant occupational health and safety guidelines while undertaking practical investigations						6.3.1					
Generate, collate and record data	systematically generate and record primary data, and collate secondary data, appropriate to the investigation						6.3.1					11.2.1
	record and summarise both qualitative and quantitative data, including use of a logbook as an authentication of generated or collated data						6.3.1					11.2.1
	organise and present data in useful and meaningful ways, including tables, bar charts and line graphs	1.4.1					6.1.2 6.1.3 6.2.1 6.2.2 6.3.1				10.1.2 10.1.3	11.2.1

Analyse and evaluate data and investigation methods	process quantitative data using appropriate mathematical relationships and units, including calculations of percentages, percentage change and measures of central tendencies (mean, median, mode), and demonstrate an understanding of standard deviation as a measure of variability	1.4.1 1.4.2									
	identify and analyse experimental data qualitatively, applying where appropriate concepts of: accuracy, precision, repeatability, reproducibility and validity; errors; and certainty in data, including effects of sample size on the quality of data obtained	1.4.3				6.2.3 6.3.1		8.2.4			11.1.2 11.2.2
	identify outliers and contradictory or incomplete data	1.4.2				6.3.1					
	evaluate investigation methods and possible sources of error or uncertainty, and suggest improvements to increase validity and to reduce uncertainty	1.4.3				6.2.3 6.3.1	7.3.2	8.2.4		10.1.4	11.1.2 11.1.3 11.1.4 11.2.1 11.2.2 11.3.1
Construct evidence-based arguments and draw conclusions	distinguish between opinion, anecdote and evidence, and scientific and non-scientific ideas	1.4.4									
	evaluate data to determine the degree to which the evidence supports the aim of the investigation, and make recommendations, as appropriate, for modifying or extending the investigation	1.4.5				6.2.3 6.3.1		8.2.4			11.2.2
	evaluate data to determine the degree to which the evidence supports or refutes the initial prediction or hypothesis	1.4.5				6.2.3 6.3.1					11.2.2
	use reasoning to construct scientific arguments, and to draw and justify conclusions consistent with evidence base and relevant to the question under investigation	1.4.5	2.1.3	3.5.2	4.3.2	5.5.3	6.2.3 6.3.1	7.2.4	8.2.4		11.2.2
	identify, describe and explain the limitations of conclusions, including identification of further evidence required	1.4.5				6.2.3 6.3.1		8.2.4			11.2.2
	discuss the implications of research findings and proposals, including appropriateness and application of data to different cultural groups and cultural biases in data and conclusions	1.4.5	2.1.3	3.5.2		5.5.3					

Analyse, evaluate and communicate scientific ideas	use appropriate psychological terminology, representations and conventions, including standard abbreviations, graphing conventions and units of measurement	1.1.1 1.1.3 1.3.1 1.4.7	2.1.1 2.2.1 2.3.2	3.1.1	4.2.1 4.2.2 4.2.3 4.2.4 4.2.5 4.2.6 4.3.1	5.1.1	6.3.1	7.2.4				
	discuss relevant psychological information, ideas, concepts, theories and models and the connections between them	1.1.1 1.1.3 1.3.1	2.1.1 2.1.2 2.2.1 2.3.1 2.3.3 2.4.1	3.1.1 3.1.2 3.1.3 3.1.4 3.2.1 3.3.1 3.4.1 3.4.2 3.5.1 3.5.2	4.1.1 4.2.1 4.2.2 4.2.3 4.2.4 4.2.5 4.2.6 4.3.1 4.4.1	5.1.1 5.2.1 5.2.2 5.3.1 5.4.1 5.5.1 5.5.2 5.5.3	6.1.1 6.1.2 6.1.3 6.2.1 6.2.2 6.2.3 6.3.1 6.3.2 6.3.3	7.1.1 7.1.2 7.2.1 7.2.2 7.2.3 7.3.3 7.3.4	8.1.1 8.1.2 8.1.3 8.1.4 8.2.1 8.2.2 8.2.3 8.2.4 8.2.5	9.1.1 9.1.2 9.1.3 9.1.4 9.2.1 9.2.2 9.2.3 9.2.4 9.2.5	10.1.1 10.1.2 10.1.3 10.1.4 10.1.5 10.2.1 10.2.2 10.2.3	11.1.3 11.1.4 11.1.5 11.2.2 11.3.1 11.3.2
	analyse and explain how models and theories are used to organise and understand observed phenomena and concepts related to psychology, identifying limitations of selected models/theories				4.2.1 4.2.2 4.2.6 4.3.1	5.2.1		7.2.2		9.1.2 9.1.4 9.2.2 9.2.5		
	critically evaluate and interpret a range of scientific and media texts (including journal articles, mass media communications, opinions, policy documents and reports in the public domain), processes, claims and conclusions related to psychology by considering the quality of available evidence				4.2.7	5.4.1						
	analyse and evaluate psychological issues using relevant ethical concepts and principles, including the influence of social, economic, legal and political factors relevant to the selected issue									9.2.5		
	use clear, coherent and concise expression to communicate to specific audiences and for specific purposes in appropriate scientific genres, including scientific reports and posters							7.3.1				
	acknowledge sources of information and assistance, and use standard scientific referencing conventions					5.4.1						

1 Scientific research methods

1.1 The process of psychological research investigations

1.1.1 Developing your research question, aim and hypothesis

> **Key science skills**
> Develop aims and questions, formulate hypotheses and make predictions
> - identify, research and construct aims and questions for investigation
> - formulate hypotheses to focus investigation
>
> Analyse, evaluate and communicate scientific ideas
> - use appropriate psychological terminology, representations and conventions, including standard abbreviations, graphing conventions and units of measurement
> - discuss relevant psychological information, ideas, concepts, theories and models and the connections between them

Develop

Understanding and being able to use key terms correctly in Psychology is important for success in this subject. This activity is structured to assist you in finding the definitions of some key terms used in science investigation, and then applying these key terms to different scenarios.

PART A

1 Define:

  a Research question

  _____

  _____

  b Aim

  _____

  _____

  c Hypothesis

  _____

  _____

This activity is testing to see if you understand the difference between a research question, aim and hypothesis.

PART B

Below are three ideas that a Year 12 student, Alexi, has been considering for her research investigation. For each idea presented:
- construct a research question
- write an aim
- formulate a hypothesis.

IDEA 1: USING REWARDS TO CHANGE BEHAVIOUR IN FIVE-YEAR-OLDS

Alexi has five-year-old twin nieces who she babysits every Wednesday and Thursday after school. After learning about operant conditioning in her Psychology class, Alexi wants to find out if the use of positive reinforcement (reward) would make it easier for her to get her nieces to tidy up their toys at night. She is thinking about using a reward such as a lolly with one niece but not with the other.

1 Construct a research question.

2 Write an aim.

3 Formulate a hypothesis.

IDEA 2: USING MNEMONICS TO IMPROVE MEMORY RETRIEVAL

Alexi is finding it difficult to remember all the information she is learning in her Psychology class. She is keen to find out if the use of mnemonics will help her to remember more material. A group of six students in her class have agreed to help her with her investigation.

1 Construct a research question.

2 Write an aim.

3 Formulate a hypothesis.

IDEA 3: EFFECT OF PARTIAL SLEEP DEPRIVATION ON LEARNING

Alexi has a job at a restaurant on Friday and Saturday nights. She often does not get home until after midnight and does not fall asleep until 2 a.m. She then finds it difficult to study on Sunday as she is very tired. Alexi is wondering if the lack of sleep is impacting her grades and wants to investigate this. A group of four students in her class have agreed to help her with her investigation.

1 Construct a research question.

2 Write an aim.

3 Formulate a hypothesis.

1.1.2 Predicting outcomes

Key science skills
Develop aims and questions, formulate hypotheses and make predictions
- predict possible outcomes of investigations

A hypothesis is a statement of what you think the direction of results in a scientific investigation will be. A hypothesis can be supported by the data collected or it can be refuted (not supported). For the following hypotheses, state the predicted outcome if the hypothesis is:
- supported by the data collected by the investigation
- refuted by the data collected by the investigation.

HYPOTHESIS 1: SECONDARY STUDENTS WHO EAT BREAKFAST BEFORE SCHOOL WILL GET BETTER GRADES ON THEIR ENGLISH EXAM THAN SECONDARY STUDENTS WHO DO NOT EAT BREAKFAST BEFORE SCHOOL.

1 State the predicted outcome if the hypothesis is:

  a supported

  b refuted

HYPOTHESIS 2: VCE STUDENTS WHO SUFFER FROM ANXIETY IN THE MONTH BEFORE AN EXAM WILL GET HIGHER GRADES THAN VCE STUDENTS WHO DO NOT SUFFER FROM ANXIETY IN THE MONTH BEFORE AN EXAM.

1 State the predicted outcome if the hypothesis is:

  a supported

  b refuted

HYPOTHESIS 3: P-PLATE DRIVERS WHO ARE DISTRACTED BY A PHONE CALL WILL BE MORE LIKELY TO MAKE ERRORS ON A DRIVING SIMULATOR THAN THOSE WHO ARE NOT DISTRACTED.

1 State the predicted outcome if the hypothesis is:

  a supported

  b refuted

HYPOTHESIS 4: YEAR 12 STUDENTS WILL HAVE GREATER RECALL OF A LIST OF 20 WORDS IF THEY ARE TAUGHT A MNEMONIC TECHNIQUE INVOLVING IMAGERY COMPARED WITH STUDENTS WHO ARE NOT TAUGHT A MNEMONIC TECHNIQUE.

1 State the predicted outcome if the hypothesis is:

a supported

....

....

b refuted

....

....

1.1.3 Sampling techniques

Key science skills

Plan and conduct investigations
- design and conduct investigations; select and use methods appropriate to the investigation, including consideration of sampling technique (random and stratified) and size to achieve representativeness, equipment and procedures, taking into account potential sources of error and uncertainty; determine the type and amount of qualitative and/or quantitative data to be generated or collated

Analyse, evaluate and communicate scientific ideas
- use appropriate psychological terminology, representations and conventions, including standard abbreviations, graphing conventions and units of measurement
- discuss relevant psychological information, ideas, concepts, theories and models and the connections between them

Develop

Sampling is an important concept in psychological investigations. You can't test everyone in the world, your town or your school, so you must select a sample. There are different ways of doing this.

1 Let's see what you know about sampling. Fill in the missing words in the sentence below using these words: chance, population, participants, sampling, representative.

_____ is the process of selecting a _____ group of people from a nominated _____. A group will be representative if every member of the population has equal _____ of being selected to participate. The people who do end up being selected in the sample are then called _____.

2 Table 1.1 shows three different sampling techniques. Complete this table by naming each technique, describing each technique, then listing its benefits and limitations.

Table 1.1 Description and evaluation of sampling techniques

Sampling technique	Name of technique	Description of technique	Benefits of technique	Limitations of technique
Figure 1.1				

Sampling technique	Name of technique	Description of technique	Benefits of technique	Limitations of technique
Figure 1.2				
Figure 1.3				

3 Name the sampling technique that would be best to use in each of the following investigations.

  a The prevalence of phobia in the adult population of a large town in Victoria.

   Explain your choice: _____

  b Subject preferences of secondary school students in a suburban high school.

   Explain your choice: _____

  c Conducting a happiness survey of 50 workers in a company that employs 200 people.

   Explain your choice: _____

  d Finding out the food preference of 10 people as a homework task.

   Explain your choice: _____

1.2 Scientific investigation methodologies

1.2.1 Methodologies

Key science skills
Plan and conduct investigations
- determine appropriate investigation methodology: case study; classification and identification; controlled experiment (within subjects, between subjects, mixed design); correlational study; fieldwork; literature review; modelling; product, process or system development; simulation

Practise

The methodology is the broader framework of the approach that is taken in the investigation. Table 1.2 presents a number of different research aims. Your task is to match the aim of each investigation with the best methodology that could be used to study it. Write your choice of methodology in the first column.

Methodologies to choose from:
- case study
- fieldwork
- literature review
- classification and identification
- modelling
- correlational study
- simulation
- controlled experiment.

Table 1.2 Research aims and appropriate methodologies

Methodology	Research aim
	To evaluate published research that has reported successful use of operant conditioning to train children to go to sleep.
	To construct a small cardboard replica of an Ames Room in order to understand how it works.
	To observe which shelves adults and children look at first when shopping in a supermarket.
	To input data into a computer program about the behaviour of bystanders at an accident.
	To establish whether the use of acrostics assists in the retention of word lists.
	To group people with similar behavioural symptoms and then use these symptoms to provide a name for the group, using the DSM-5.
	To investigate the use of different bedtime rituals to help people get to sleep.
	To explore the life of a child diagnosed with ADHD by documenting their unique behaviours over a period of four weeks.
	To find out whether levels of anxiety increase as income level decreases.

1.3 The controlled experiment in detail

1.3.1 Variables in controlled experiments

Key science skills
Develop aims and questions, formulate hypotheses and make predictions
- identify independent, dependent and controlled variables in controlled experiments

Analyse, evaluate and communicate scientific ideas
- use appropriate psychological terminology, representations and conventions, including standard abbreviations, graphing conventions and units of measurement
- discuss relevant psychological information, ideas, concepts, theories and models and the connections between them

Understanding and being able to use key terms correctly in Psychology is important for success in this subject. This activity is structured to assist you in finding the definitions of some key terms used in science investigation, and then applying these key terms to different scenarios.

PART A

1 Define:

a Independent variable

_____

b Dependent variable

_____

c   Controlled variable

d   Extraneous variable

e   Confounding variable

> This activity is testing to see if you understand what a variable is and the different types of variables.

PART B

Refer back to the three different ideas that Alexi had for her research investigation in Activity 1.1.1. For each of these ideas, state:
- the independent variable
- the dependent variable
- any potential extraneous variables that she will need to control
- any potential confounding variables.

IDEA 1: USING REWARDS TO CHANGE BEHAVIOUR IN FIVE-YEAR-OLDS

1   State:

a   the independent variable

b   the dependent variable

c   the controlled variable

d   potential extraneous variables that need to be controlled

e   potential confounding variables

IDEA 2: USING MNEMONICS TO IMPROVE MEMORY RETRIEVAL

1   State:

a   the independent variable

b   the dependent variable

c   the controlled variable

d potential extraneous variables that need to be controlled

e potential confounding variables

IDEA 3: EFFECT OF PARTIAL SLEEP DEPRIVATION ON LEARNING

1 State:

   a the independent variable

   b the dependent variable

   c the controlled variable

   d potential extraneous variables that need to be controlled

   e potential confounding variables

1.3.2 Experimental designs for controlled experiments

Key science skills
Plan and conduct investigations
- determine appropriate investigation methodology: case study; classification and identification; controlled experiment (within subjects, between subjects, mixed design); correlational study; fieldwork; literature review; modelling; product, process or system development; simulation

If you are using the controlled experiment methodology to conduct an investigation, you need to decide if you are going to use a within subjects, between subjects or mixed design. To understand the difference between these designs, complete the following activity.

1 Define the following three types of experimental design:

   a within subjects

   b between subjects

   c mixed design

2   In Table 1.3, choose which of the three experimental designs would be best for each investigation. Write your choice of experimental design in the second column.

Table 1.3 Investigations and experimental designs

Investigation	Design
Half the sample receive a drug that is believed to improve memory and the other half of the sample receive a placebo. Their retention of a list of 20 words is tested 10, 60 and 120 minutes later.	
40 participants were each shown 20 photos in random order: 10 photos were of kittens and 10 photos were of adult cats. The dilation of the participants' pupils was measured as they viewed each photo.	
To test reaction time, 12 elderly people were asked to press a button when a light came on. Each person was tested three times: in the morning, at noon and just before bedtime. Each time they were tested, the light was presented 20 times and their reaction time was then averaged.	
An investigation into the calming effects of music saw 10 VCE students placed into a group that listened to 15 minutes of classical music. Three days later, the same 10 students listened to 15 minutes of jazz music. A separate group of 10 students listened to no music. Students were then immediately tested using an online anxiety test.	
To test motivation levels, 50 primary school students were taught a short poem. Half the children received rewards (lollies) during their lesson if they got at least half of the poem correct; the other half did not. At the end of the lesson the students' ability to recite the poem correctly was tested.	

1.4 Analysing and evaluating research

1.4.1 Displaying data in tables, bar charts and line graphs

Key science skills
Generate, collate and record data
- organise and present data in useful and meaningful ways, including tables, bar charts and line graphs

Analyse and evaluate data and investigation methods
- process quantitative data using appropriate mathematical relationships and units, including calculations of percentages, percentage change and measures of central tendencies (mean, median, mode), and demonstrate an understanding of standard deviation as a measure of variability

When psychological research is conducted, raw data is collected. This is data that has not been organised or displayed in any way. In this activity you will practise organising raw data and displaying it in the most appropriate way.

Arryn and Sean conducted a research investigation to find out the levels of stress experienced by students in Years 10–12. They randomly sampled 10 students from each year level at their school. Each participant completed the Sharrock online stress test, which generates a score out of 50. The higher the score, the higher the stress level being experienced by the participant. The raw data from this investigation is presented in Figure 1.4.

Figure 1.4 Raw data from stress level investigation

Year 10: 27, 32, 21, 28, 31, 20, 29, 22, 26, 20
Year 11: 31, 30, 34, 29, 25, 27, 29, 33, 27, 25
Year 12: 34, 33, 41, 40, 39, 36, 34, 42, 45, 37

1 Use the space below to show how this raw data could best be organised in your logbook. Include any measures of central tendency that you think are useful.

2 Now use the graph paper below to construct a graph that will best display this data. Think about how you can display the data so people reading your report can see immediately what your results show.

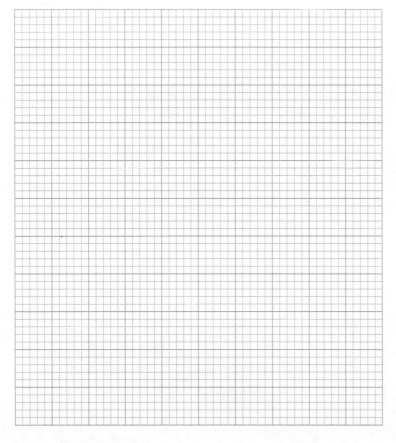

Remember to include all the features of a well-drawn graph. Refer back to your textbook if you are not sure what these are.

1.4.2 Descriptive statistics

Key science skills
Analyse and evaluate data and investigation methods
- process quantitative data using appropriate mathematical relationships and units, including calculations of percentages, percentage change and measures of central tendencies (mean, median, mode), and demonstrate an understanding of standard deviation as a measure of variability
- identify outliers and contradictory or incomplete data

Table 1.4 provides the raw data from an investigation in which 11 participants received positive feedback while learning a 20-word list. Two days later, the same participants received negative feedback while learning another 20-word list. Before commencing, the words on each list were matched for word length and difficulty. Study the table to make sure that you understand what it is telling you, and then answer the questions that follow.

Table 1.4 Raw data from feedback investigation

Participant number	Number of words remembered in a 20-word list		Percentage change (%)
	Negative feedback	Positive feedback	
1	8	17	
2	11	15	
3	12	11	
4	6	16	
5	12	15	
6	12	17	
7	10	12	
8	9	18	
9	4	17	
10	8	10	
11	7	14	
	Mean =	Mean =	
	Median =	Median =	
	Mode =	Mode =	

1. For each condition in this investigation, calculate the mean, median and mode. Write your answers in the appropriate boxes at the bottom of Table 1.4.

2. Complete the last column of Table 1.4 by calculating the percentage change in scores for each participant. (Note: one of the answers will be negative.)

3. One set of data could be considered an outlier. What is the number of that participant?

CHAPTER 1 / Scientific research methods

1.4.3 Quality of data

Key science skills
Analyse and evaluate data and investigation methods
- identify and analyse experimental data qualitatively, applying where appropriate concepts of accuracy, precision, repeatability, reproducibility and validity; errors; and certainty in data, including effects of sample size on the quality of data obtained
- evaluate investigation methods and possible sources of error or uncertainty, and suggest improvements to increase validity and to reduce uncertainty

Develop

Maria plays basketball and wants to ensure that she is the best in her school at three-point shots. In order to do this, her coach told her that she needed to be precise in her shooting ability, while her brother told her that she needed to be accurate in her shots. She was very confused with these two pieces of advice and didn't think there was any difference between the two. Maria then remembered that these terms were also used by her Psychology teacher when it came to scientific investigations. Her Psychology teacher drew four dartboards (see Figure 1.5) and used them to help explain the difference between accuracy and precision.

Accuracy refers to how close the value is to the true value, while precision refers to how close repeated values are to each other.

1. Figure 1.5(a) shows low accuracy and low precision. Complete Figure 1.5(b–d) by drawing darts that show the following precision and accuracy levels.
   a. low accuracy, low precision
   b. high precision, low accuracy
   c. low precision, high accuracy
   d. high precision, high accuracy

Figure 1.5

 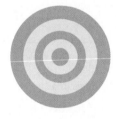  

A   B   C   D

2. Complete Table 1.5 by filling in the missing information.

Table 1.5 How to ensure data quality

What do I have to watch out for with my data?	What exactly is this?	What do I have to do to ensure the quality of my data in relation to this feature?
	How close the value is to the true value of the quantity being measured.	
Precision		Make sure you use measuring equipment properly and have a large sample size. Focus on what you are doing to make sure there are no random errors.

Repeatability		
	The same quantity being measured by a different person under different conditions.	
Validity		
	Measurements that lie a long way from other measurements.	Make sure you measure carefully and that equipment is working properly. Can sometimes be removed from data before analysis.
Sample size		Have a sample that is representative of the population.
Internal validity		
	The results of the investigation can be applied to similar individuals in a different setting.	
	Unpredictable variations in the measurement process that affect precision.	
Systematic error		
		Focus! Double check all measurements and recordings.
	Lack of exact knowledge of the value of the quantity being measured.	

1.4.4 Know what you are reading!

> **Key science skills**
> Construct evidence-based arguments and draw conclusions
> - distinguish between opinion, anecdote and evidence, and scientific and non-scientific ideas

Read the following passage carefully, taking note of the types of information you are reading, then answer the questions that follow.

Can you always believe what you read?

A recurring argument that my best friend, Marli, and I have is about how many 'other' friends we should have. From what she has been saying and doing, I think Marli has been questioning this a lot during the COVID-19 pandemic.

We both already have one or two 'other' friends, as well as brothers, sisters and cousins we like to spend time with. I think I am a bit of an introvert, so this feels like plenty of friends to me. In fact, it could be too many! But I think that Marli is definitely an extrovert. She comes alive around other people and loves to party. I am noticing that as we slowly come out of the pandemic, Marli is making noises about wanting to reach out and find more friends to join our little group.

From what I have read online, it is not unusual to want a small group of friends and then lots of people who are not really close friends. I have read that for years friendship in Australia has been getting worse. An article in *Molly Magazine* said that three decades ago, 4 per cent of Australians had no close friends at all! Imagine that! I guess they didn't have things like mobile phones and social media back then. An online poll in 2022, *What do you think?*, produced a result closer to 14 per cent! So, it's actually worse now!

My mum told me that when she was at school one of her teachers told her that smoking 20 cigarettes a day can be as harmful to our health as loneliness, and that is why she gave up smoking. I don't believe her; I can't imagine my mum ever smoking!

1 Define 'opinion'.

2 Find one example of an opinion in the passage.

3 Define 'anecdote'.

4 Find one example of an anecdote in the passage.

5 Define 'non-scientific ideas'.

6 Find one example of a non-scientific idea in the passage.

7 Find one conclusion made in the passage.

1.4.5 What is your data telling you?

Key science skills
Construct evidence-based arguments and draw conclusions
- evaluate data to determine the degree to which the evidence supports the aim of the investigation and make recommendations, as appropriate, for modifying or extending the investigation
- evaluate data to determine the degree to which the evidence supports or refutes the initial prediction or hypothesis
- use reasoning to construct scientific arguments and to draw and justify conclusions consistent with evidence base and relevant to the question under investigation
- identify, describe and explain the limitations of conclusions, including identification of further evidence required
- discuss the implications of research findings and proposals, including appropriateness and application of data to different cultural groups and cultural biases in data and conclusions

This activity requires you to interpret results and to draw conclusions from those results. Read the information carefully and then answer the questions that follow.

Gerry and Amanda studied the effects of positive reinforcement on students' performance on a grammar exam. They used a sample from Year 9 at Valley High School, which consisted of 80 students: 40 males and 40 females aged between 14 and 15 years old. These students were spread over three English classes (Class D, Class E and Class F). The distribution of students across these three classes was:

- Class D: 16 males; 9 females
- Class E: 15 males; 15 females
- Class F: 9 males; 16 females

Each student completed a grammar test (Test 1), marked out of 50, at the start of the investigation.
The teachers of each class were instructed to:

- Class D: continue teaching as normal
- Class E: give students a sticker when they completed a section of work
- Class F: give students a sticker and 10 minutes' free time when they completed a section of work.

At the end of five weeks, the students were given another grammar test (Test 2), also marked out of 50.
The results of both tests for each class are shown in Table 1.6.

Table 1.6 Grammar results for Year 9 students, Valley High School

Class	Mean score on Test 1	Mean score on Test 2	Percentage difference
Class D	29.1	30.4	4.47
Class E	27.3	33.5	22.7
Class F	28.6	37.8	32.2

1. Write an aim and a hypothesis for this study.
   a. Aim: _____
   b. Hypothesis: _____

2. Does the evidence presented in Table 1.6 answer the aim? Explain your response.
_____
_____
_____

3 Does the evidence presented in Table 1.6 support or refute the hypothesis? Explain your answer.

4 Provide one suggestion for how this study could be extended or modified.

5 Write a conclusion for this study.

6 What are two possible limitations, caused by extraneous variables or error, that may have affected this study?

7 What are two possible implications of the findings of this study?

1.4.6 Ethics

Key science skills
Comply with safety and ethical guidelines
- demonstrate ethical conduct and apply ethical guidelines when undertaking and reporting investigations

The article below presents a psychological experiment with a number of ethical issues. Read the article carefully and answer the questions that follow.

Facebook apologises for psychological experiment on users

A Facebook executive has apologised for the conduct of secret psychological tests on nearly 700 000 users in 2012, which prompted outrage from users and experts alike.

The experiment hid 'a small percentage' of emotional words from peoples' news feeds, without their knowledge, to test what effect that had on the statuses or 'likes' that they then posted or reacted to.

'This was part of the ongoing research that companies do to test different products. That was what this study was, but it was poorly communicated,' said Facebook's chief operating officer while in New Delhi. 'And for that communication we apologise. We never meant to upset you.'

Facebook's first public comment on the experiments came as the social network attempted to woo Indian advertisers as part of its efforts to tailor adverts to users outside of the US. The aim of the government-sponsored study was to see whether positive or negative words in messages would lead to positive or negative content in status updates.

The company's researchers decided after tweaking the content of peoples' 'news feeds' that there was 'emotional contagion' across the social network, by which people who saw one emotion being expressed would then express similar emotions.

The social network faced stern criticism from commentators and researchers, over its handling of the experiment which was not explained to users and therefore was performed without their permission. Some referred to it as 'creepy', 'evil', 'terrifying' and 'super disturbing'.

Source: Adapted from Gibbs, S. (2014, July 3) Facebook apologises for psychological experiment on users. *The Guardian*.

1 Identify the aim of this study.

2 This psychological experiment breached a number of ethical guidelines. Ethical guidelines are put in place to protect an individual's rights in an experiment. Identify three ethical guidelines that have been breached by this experiment and explain how they have been breached.

a

b

c

1.4.7 Know your key terms

> **Key science skills**
> Analyse, evaluate and communicate scientific ideas
> - use appropriate psychological terminology, representations and conventions, including standard abbreviations, graphing conventions and units of measurement

Develop

1 Table 1.7 lists some of the key terms that you need to understand and be able to use. However, the key terms are not correctly matched to the definitions. Start by reading each definition, then find a key term that matches the definition. Write the letter that is to the left of the key term in the answer column.

Table 1.7 Key terms

	Key term	Definition	Key term's letter
a	Personal error	How close a measurement is to the 'true' value of the quantity being measured.	
b	Internal validity	How closely a set of measurement values obtained for a behaviour, psychological construct or trait of interest are to one another.	
c	Accuracy	How closely a set of measurements for a particular behaviour, psychological construct or trait match each other when the measurement is done again by the same researcher in the same laboratory using the same procedure and measurement instrument on the same sample or individual, within a short period of time.	
d	Controlled variable	The extent of similarity between measurements obtained for a particular behaviour or psychological construct or trait when the measurements are conducted by different researchers, located in different research settings, using different samples of participants.	
e	Confidentiality	The value, or range of values, that would be found if the quantity could be measured perfectly.	
f	Dependent variable	The extent to which the design of a scientific investigation and the measurements chosen provide meaningful and generalisable information about the psychological constructs of interest.	
g	Beneficence	When the study design and methods include effective measures of the psychological constructs of interest so that the study's results can be interpreted meaningfully in relation to the aims of the study.	
h	Extraneous variable	When the results can be generalised to relevant populations beyond the sample.	
i	Precision	Commitment to maximising benefits and minimising risks involved in a study.	
j	Independent variable	Ensuring there is a fair consideration of competing claims.	
k	Systematic errors	Protection and privacy of a participant's personal information.	
l	True value	Errors that affect the precision of a measurement.	
m	Random errors	Errors that affect the accuracy of a measurement.	
n	Repeatability	Mistakes made by the person conducting the research.	
o	Reproducibility	Variables that are held constant by the researcher.	
p	Justice	Variable that is manipulated during research.	
q	Validity	Variable that is measured during research.	
r	External validity	Variable that may affect the results of an investigation and needs to become a controlled variable.	

Exam practice

Multiple choice

Circle the response that best answers the question.

1. Which of the following references is written in correct APA style?
   A. Peter Smith (2021). The Journal of sleep, Identifying sleep patterns in felines, volume 3, pages 2–34
   B. Naughty and Nice, The Journal of Bad behaviour, written by Carole Norton in 2018
   C. Carbohydrates in Pregnancy, West and Wayner, 2018, Volume 1, pages 22–24
   D. Christopher, P., & Blackwood, T. (2020). Orthodontic applications in pregnant newts. *Journal of Dentistry, 1*, 1–11

2. Participants who are used in research because they were available, close by or easily recruited are known as a
   A. random sample.
   B. stratified sample.
   C. sample of convenience.
   D. non-standardised sample.

3. When conducting experimental research, it is intended that the experiment will identify the effect of the
   A. dependent variable.
   B. independent variable.
   C. extraneous variables.
   D. confounding variables.

4. In a study investigating the effect of alcohol on driving performance, participants are asked to drink alcoholic drinks of varying concentrations before being tested in a driving simulator. Participants who are in the control group in this experiment are likely to drink
   A. a non-alcoholic drink.
   B. an alcoholic drink with the lowest concentration of alcohol before being tested.
   C. an alcoholic drink with the highest concentration of alcohol before being tested.
   D. both alcoholic and non-alcoholic drinks.

5. An experiment investigated how sleep deprivation affects memory. Participants were kept awake for periods of time ranging from three hours to nine hours. At the end of the period, all participants were given a learning task and were tested on their recall. The independent variable was the _____. The dependent variable was the _____.
   A. number of hours of sleep deprivation; score on the recall test
   B. score on the recall test; number of hours of sleep deprivation
   C. presence of two groups; learning task
   D. learning task; presence of two groups

6. One reason why an experimenter would use deception in research is that
   A. the experiment will cause some distress and the experimenter does not want the participant to withdraw.
   B. the participant is a university or school student studying Psychology and is therefore not naive.
   C. the research is testing morals or values that could be embarrassing or upsetting.
   D. if participants have too much information they will change their way of behaving, which may influence the results.

7. Psychological research aims to use a random sample when conducting research because
   A. it provides a broad range of responses from participants.
   B. it ensures that the sample is representative of the population.
   C. it helps eliminate any experimenter effects.
   D. it is always accurate.

8 Researchers randomly selected a sample of university students and gave them two energy drinks a day to investigate the effect they had on depression. The research showed that people who consume energy drinks on a daily basis are more likely to report depressive symptoms compared to those who do not drink energy drinks. This study is an example of a
   A controlled experiment.
   B correlational study.
   C literature review.
   D simulation.

9 Researchers want to test whether 'absence makes the heart grow fonder'. Their participants will be couples who are separated for six months due to work commitments. They plan to periodically have participants write responses to sets of statements about their feelings. What steps should the researchers follow in organising the research?
   A obtain a group of people prepared to act as participants, identify the research problem, design the method, formulate a testable hypothesis
   B identify the research problem, obtain a group of people prepared to act as participants, design the method, formulate a testable hypothesis
   C formulate a testable hypothesis, obtain a group of people prepared to act as participants, identify the research problem, design the method
   D identify the research problem, formulate a testable hypothesis, design the method, obtain a group of people prepared to act as participants

10 Graphs are a convenient way of presenting data. In a research report they are most likely to be found in the
   A introduction.
   B method.
   C results.
   D discussion.

11 Once research has been conducted, researchers prepare a report of their findings and present it to others, either at a conference or by publishing it in a journal. The purpose of such reporting is to
   A enable other psychologists to benefit from the newly acquired knowledge.
   B allow other psychologists to replicate the study.
   C enable the general public to learn from the research.
   D all of the above.

12 Mitchell is conducting an experiment to test the idea that people perceived to be attractive are judged by others to be more honest than unattractive people. Which of the alternatives below is the best statement of his hypothesis?
   A Attractive people are more honest than unattractive people.
   B A person's attractiveness will influence others' opinions of him.
   C Will attractive people be judged as more honest than unattractive people?
   D Individuals judged to be attractive will be more likely to be considered honest than individuals judged to be unattractive.

13 It is important to obtain informed consent when conducting psychological research. Informed consent involves
   A participants volunteering to be a part of the study.
   B participants signing an agreement that outlines the details of the study and confirms their involvement.
   C participants being free to leave the study at any time.
   D participants' private information being kept confidential at all times.

14 Ms Vu is giving her class a test on fractions at the end of the week. She is very proud of the difficulty of the questions she has written, as they all require the students to use logic and analytical skills to solve the mathematical problems. Her colleagues feel that her test does not actually assess ability to manipulate fractions, but reasoning skills instead, therefore her test does not have
   A generalisability.
   B reliability.
   C validity.
   D tangibility.

Short answer

Use the following information to answer Questions 1–6.

Figure 1.6 shows the results of an investigation involving inhabitants of a small rural town in Victoria. The aim of the investigation was to find who was using the library during the day. The graph shows the age of people who visited the local library in a one-hour period.

Figure 1.6

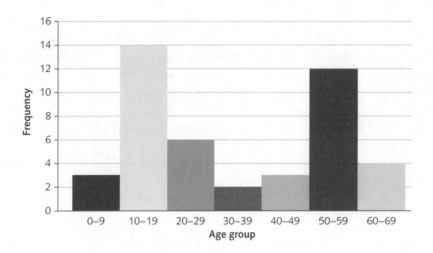

1 How many people visited the library in this one-hour period?   1 mark

2 Which two groups of people attended the library the most during this one-hour period?   1 mark

3 Which group of people attended the library the least during this one-hour period?   1 mark

4 What methodology was used in this investigation?   1 mark

5 What significant limitation is highlighted by the method used in this study?   1 mark

6 Comment on the generalisability of this study.   2 marks

Nervous system functioning 2

2.1 The nervous system: roles and subdivisions

Key knowledge
- the roles of different subdivisions of the central and peripheral nervous systems in responding to, and processing and coordinating with, sensory stimuli received by the body to enable conscious and unconscious responses, including spinal reflexes

2.1.1 Nervous system 'What am I?'

Key science skills
Analyse, evaluate and communicate scientific ideas
- use appropriate psychological terminology, representations and conventions, including standard abbreviations, graphing conventions and units of measurement
- discuss relevant psychological information, ideas, concepts, theories and models and the connections between them

Develop

In this activity you will revise your understanding of the divisions of the nervous system and the functions they perform.

PART A

1 Draw the divisions of the central nervous system inside the outline of the person in Figure 2.1. Clearly label these biological structures and write bullet points to identify their role in the transmission of information around the body. The bullet points should be beside the outline.
2 Using a different colour, draw in the peripheral nervous system and clearly label it. Write bullet points to identify the role of the PNS in the transmission of information around the body. The bullet points should be beside the outline.

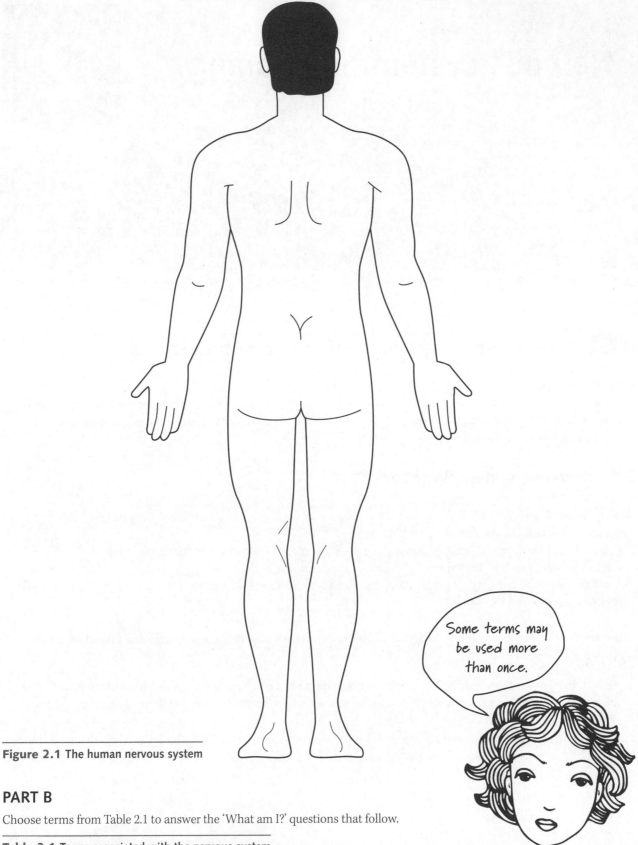

Figure 2.1 The human nervous system

PART B

Choose terms from Table 2.1 to answer the 'What am I?' questions that follow.

Table 2.1 Terms associated with the nervous system

peripheral nervous system	sensory information	spinal cord
parasympathetic nervous system	somatic nervous system	sympathetic nervous system
motor information	flight-or-fight-or-freeze response	autonomic nervous system
central nervous system	brain	

Some terms may be used more than once.

'WHAT AM I?' QUESTIONS

1. If I lost control of the muscles attached to your skeleton, you would not be able to move. What am I?

2. I operate like an information highway. I carry the sensory information brought to me by the peripheral nervous system to the brain for processing and interpretation. I also carry instructions about movement from the brain to the peripheral nervous system. What am I?

3. Although at times you can consciously influence some of my activities, I usually operate on my own to regulate and modify your internal functioning, including your heart rate, digestion rate and glandular activity. What am I?

4. When you feel threatened, I am automatically triggered by the activation of the sympathetic nervous system. I let you choose between facing the threat or removing yourself from its presence. What am I?

5. My effect begins when a motor message is sent to your adrenal glands, telling them to release more adrenaline and noradrenaline into your bloodstream. I also trigger an increase in the levels of fats and sugars released into your bloodstream, altering your internal functioning so your body has more energy. You know I am at work because your mouth feels dry, you start to perspire and you feel your bladder relaxing. What am I?

6. My nerve network receives sensory information from the peripheral nervous system, integrates and interprets it, and delivers motor messages to the peripheral nervous system. What am I?

7. In most people I dominate for the majority of time. I ensure that your internal systems related to survival are operating at a balanced level so you enjoy health and wellbeing. What am I?

8. I am composed of all the nerves that do not reside in your brain or spinal cord and I have a sensory and a motor function. You depend on me to control all your voluntary actions as well as the involuntary functions of your internal organs, muscles and glands. What am I?

9. I am connected to the rest of the body by an intricate and delicate cable of nerve fibres that brings me sensory information. I process this information, integrate it and decide on appropriate responses to it. What am I?

10. When you need extra energy, I alter the activity level of your visceral muscles, organs and glands so that your body can produce this extra energy. When you no longer need the extra energy, I return your visceral muscles, organs and glands to their normal level of activity and keep them working at this level. What am I?

11. I am the information that the somatic nervous system receives from sensory receptors located throughout the body and transmits to the central nervous system. What am I?

12. I am the body's automatic response to a perceived threat and I am accompanied by automatic physiological changes that prepare the body to deal with the threat. What am I?

13. Without me controlling the muscles attached to your skeleton, you would not be able to move whenever you wanted. What am I?

14. When the somatic nervous system delivers me to appropriate body parts, they move the way the brain decides they will move. What am I?

15. I can be subdivided into two distinct nervous systems. One of my subdivisions controls the voluntary movement of skeletal muscles while the other subdivision regulates the activity level of the body's internal organs and glands. What am I?

PART C

1 Identify the division of the nervous system responsible for the control of each activity.

Activity	Division of the nervous system responsible for control
Regulating your perspiration rate when you are relaxed	
Formulating instructions that will be sent to your hand when you want to grip the fridge door to open it	
Taking motor commands from the spinal cord to your hand so you can tie your shoelaces	
Transmitting an instruction from the brain to your peripheral nervous system	
Dilating your pupils when you experience fear	
Differentiating between different types of sensory stimuli	
Regulating the level of blood glucose during times of arousal and calm	
Receiving sensory information from the peripheral nervous system, interpreting it, deciding on an appropriate response and carrying motor messages back to the peripheral nervous system	
Relaying sensory information to, and motor information from, the central nervous system	
Triggering the flight-or-fight-or-freeze response	
Receiving and processing sensory information transmitted via the spinal cord and deciding on the appropriate response to this information	
Diverting extra energy to your muscles when you are swimming away from something scary	
Relaying information to the smooth muscles that control the activity levels of your internal organs and glands	
Increasing the contraction rate of your stomach after you have stopped swimming so that your digestion rate becomes normal	
Receiving sensory information brought in by the peripheral nervous system, conveying this information to the brain and transmitting motor messages from the brain to the peripheral nervous system	
Planning your next holiday overseas	
Regulating your breathing rate when you are relaxed	
Regulating your energy level during your day at school	
Energising or calming your body according to what you are thinking about	
Transmitting the sensation of a tickle on your foot to your central nervous system	

2.1.2 Nervous system concept map

Key science skills
Analyse, evaluate and communicate scientific ideas
- discuss relevant psychological information, ideas, concepts, theories and models and the connections between them

Develop

This activity will strengthen your understanding of the hierarchical structure of the human nervous system. You could challenge yourself to complete this task without referring to your notes. How much can you do?

What you need

Scissors, glue, A3 sheet of paper, highlighters

1. Each of the boxes in Table 2.2 contains the name of a division of the human nervous system. Carefully cut out these boxes.
2. Arrange these boxes on your sheet of A3 paper so that they represent a concept map of the divisions of the nervous system in hierarchical order. Do not glue them to the paper yet. Make sure that you don't crowd your boxes – spread the divisions out over the page.
3. Each of the boxes in Table 2.3 contains information relating to a specific division of the nervous system. Carefully cut out these boxes and place them on your A3 page under the correct division names (from Table 2.2). Do not glue them to the paper yet.
4. Check your notes or your textbook and make sure that you have set your diagram out correctly. When you are certain that your information is correct, glue all the boxes onto the A3 paper and draw connecting lines to indicate the relationship between the divisions.
5. Use highlighters to colour code the divisions (for example, components of the central nervous system = blue).
6. Use your concept map as a poster. Hang it up somewhere you can see it easily so that you can study it frequently.

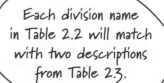

Each division name in Table 2.2 will match with two descriptions from Table 2.3.

Table 2.2 Names of divisions of the nervous system

Autonomic nervous system (ANS)	Brain	Somatic nervous system
Spinal cord	Parasympathetic nervous system	Peripheral nervous system (PNS)
Central nervous system (CNS)	Nervous system	Sympathetic nervous system

Table 2.3 Information about the divisions of the nervous system

Regulates internal bodily functions	Automatically arouses and energises the body during times of stress or need for increased physical activity	A complex communication system that is made up of billions of neurons
Transmits sensory information from internal and external environments to the central nervous system and motor commands from the central nervous system to the rest of the body	Responsible for controlling voluntary responses by receiving commands from the central nervous system and transferring this information via motor neurons to appropriate skeletal muscles	Activation triggers the flight-or-fight-or-freeze response
The 'engine room' of the nervous system	A major division of the nervous system consisting of all the nerves in the brain and spinal cord	Three major functions that include receiving, processing and organising information
Controls involuntary activity of internal muscles, organs and glands	Maintains vital functions by keeping our internal systems in a balanced and healthy state	Coordinates all the nervous system's activities
Receives sensory information from the peripheral nervous system and transmits it to the brain	Receives and interprets sensory information sent from the peripheral nervous system and transmits motor messages back to the peripheral nervous system	Transmits motor messages from the brain to the peripheral nervous system
Responsible for sensing external stimuli and transferring the information via sensory neurons to the central nervous system	Calms the body and returns internal systems to a normal level of activity when the stressor or need for increased physical activity has passed	Consists of all the nerves that extend from the brain and spinal cord throughout the body

2.1.3 Evaluation of research

> **Key science skills**
> Develop aims and questions, formulate hypotheses and make predictions
> - identify, research and construct aims and questions for investigation
> - identify independent, dependent and controlled variables in controlled experiments
> - formulate hypotheses to focus investigations
>
> Plan and conduct investigations
> - determine appropriate investigation methodology: case study; classification and identification; controlled experiment (within subjects, between subjects, mixed design); correlational study; fieldwork; literature review; modelling; product, process or system development; simulation
> - design and conduct investigations; select and use methods appropriate to the investigation, including consideration of sampling technique (random and stratified) and size to achieve representativeness, and consideration of equipment and procedures, taking into account potential sources of error and uncertainty; determine the type and amount of qualitative and/or quantitative data to be generated or collated
>
> Comply with safety and ethical guidelines
> - demonstrate ethical conduct and apply ethical guidelines when undertaking and reporting investigations
>
> Construct evidence-based arguments and draw conclusions
> - use reasoning to construct scientific arguments, and to draw and justify conclusions consistent with evidence base and relevant to the question under investigation
> - discuss the implications of research findings and proposals, including appropriateness and application of data to different cultural groups and cultural biases in data and conclusions

Develop

This activity will improve your understanding and application of research methods, while also investigating the functioning of the autonomic nervous system.

Read the research scenario and answer the questions that follow.

Can you control your autonomic nervous system?

Dr Peter Pickkers and his team from the Radboud University Nijmegen Medical Centre have investigated the effects of the autonomic nervous system on the immune response. They focused on the possibility that the autonomic nervous system and therefore the immune response can be controlled. The research was performed on a single participant, 'Iceman' Wim Hof, who claims he can influence his autonomic nervous system and immune response through concentration and meditation techniques.

To examine this, endotoxin was administered to Hof while he engaged in his concentration and meditation technique. Endotoxin is a component of bacteria and causes the body to produce an immune response that includes the production of inflammatory mediators and flu-like symptoms such as fever, chills and headaches. Different measurements were performed on Hof, including brain activity, autonomic nervous system activity and inflammatory mediators in the blood. There was an increase in the stress hormone cortisol in Hof after the administration of endotoxin. The release of cortisol usually occurs in response to increased autonomic nervous system activity, which in turn can suppress the immune response. It was found that when compared to other healthy volunteers, Hof's immune response was decreased by 50%. There were also hardly any flu-like symptoms observed in Hof.

While the results that were obtained with Hof are incredible, further research is needed that involves investigating the use of the concentration and meditation technique in one group of participants compared to another group of participants that do not use the technique.

Source: Radboud University Nijmegen Medical Centre (2011, April 22). Research on 'Iceman' Wim Hof suggests it may be possible to influence autonomic nervous system and immune response, *Science Daily*. www.sciencedaily.com/releases/2011/04/110422090203.htm

---

1 What was the aim of this investigation?

_____

_____

2 Formulate a possible hypothesis for this investigation.

_____

_____

_____

3 Identify the independent and dependent variables in this investigation.

4 Explain why this investigation would be considered a case study and not a controlled experiment.

5 Explain one benefit and one limitation of using a case study to carry out research.

6 Based on the results, what conclusion can be made about the influence of concentration and meditation techniques on the autonomic nervous system and immune response?

7 Explain one major limitation of this investigation.

8 If a controlled experiment were to be carried out for this investigation:
   a Design a procedure that could be used for this experiment.

   b Name and explain two ethical guidelines that would need to be applied by the researchers.

2.2 Conscious and unconscious responses

Key knowledge
- the roles of different subdivisions of the central and peripheral nervous systems in responding to, and processing and coordinating with, sensory stimuli received by the body to enable conscious and unconscious responses, including spinal reflexes

2.2.1 Conscious and unconscious responses of the nervous system

Key science skills
Analyse, evaluate and communicate scientific ideas
- use appropriate psychological terminology, representations and conventions, including standard abbreviations, graphing conventions and units of measurement
- discuss relevant psychological information, ideas, concepts, theories and models and the connections between them

Develop

This activity will strengthen your understanding of the difference between conscious and unconscious responses of the nervous system, including the different parts of the nervous system involved with each type of response.

PART A

1 Table 2.4 contains 10 statements about conscious and unconscious responses of the nervous system. Place a tick in the correct column to indicate whether each statement is true or false.

Table 2.4 Conscious and unconscious nervous system responses

	True	False
Voluntary responses that we make are the result of communication between the somatic nervous system and the spinal cord and operate independently of the brain.		
The brain receives and processes all types of sensory information and is then responsible for planning and executing a response.		
A spinal reflex occurs in the spinal cord, independent of the brain.		
To enable a voluntary response, information is sent from the brain via the spinal cord to motor neurons in the somatic nervous system.		
When you touch something hot you instantly withdraw your hand. This type of reflex involves sensory neurons, interneurons and motor neurons.		
A benefit of our autonomic nervous system operating automatically is that we can survive in constantly changing external environments.		
A limitation of our autonomic nervous system operating independently of the brain is that it is impossible to ever control the responses made by our internal organs and glands.		
If we are exposed to a threatening stimulus our parasympathetic nervous system is automatically activated, eliciting the emergency flight-or-fight-or-freeze response.		
Reflex actions are involuntary responses that are 'hard-wired' into our nervous system and act as a survival mechanism.		
People can use techniques such as meditation, yoga and a process known as biofeedback to gain control over the responses made by their autonomic nervous system.		

PART B

The first column of Table 2.5 contains everyday scenarios.

1. In the second column, indicate whether this is a conscious (voluntary) response or unconscious (automatic) response.
2. In the third column, explain the processes and actions of the nervous system, referring to the divisions and structures involved with each response.

Table 2.5 Nervous system responses to everyday scenarios

Scenario	Conscious or unconscious response?	Divisions and structures involved with the response
You are walking along the beach when you step on something sharp. You immediately lift your foot.		
You are responding to a text message from a friend.		
You put your hand in the bath to test the temperature of the water and decide to add more hot water.		
You put your hand in the bath to test the temperature of the water and quickly remove your hand because the water is too hot.		
You are walking in the park when a vicious dog comes charging towards you. You feel your heart race and your palms get sweaty.		
You are participating in a yoga class that requires you to engage in breathing techniques as you perform each posture.		

2.3 The transmission of neural information

Key knowledge
- the role of neurotransmitters in the transmission of neural information across a neural synapse to produce excitatory effects (as with glutamate) or inhibitory effects (as with gamma-aminobutyric acid [GABA]) as compared to neuromodulators (such as dopamine and serotonin) that have a range of effects on brain activity

2.3.1 The structure and function of neurons

Key science skills
Analyse, evaluate and communicate scientific ideas
- discuss relevant psychological information, ideas, concepts, theories and models and the connections between them

Develop

This activity will revise your understanding of the structure and function of neurons and where neurotransmitters are produced and act.

PART A

1. Use the following terms to label Figure 2.2, by writing the label and then drawing an arrow to indicate where each structural component is located.
    - Soma
    - Axon
    - Dendrites
    - Synaptic cleft (can also be called synaptic gap)
    - Axon terminal
2. Draw a circle on Figure 2.2 to show where neurotransmitters flow from one neuron to the next.
3. Draw an arrow on Figure 2.2 to show the direction of the flow of information along the neuron.

Figure 2.2 A typical motor neuron

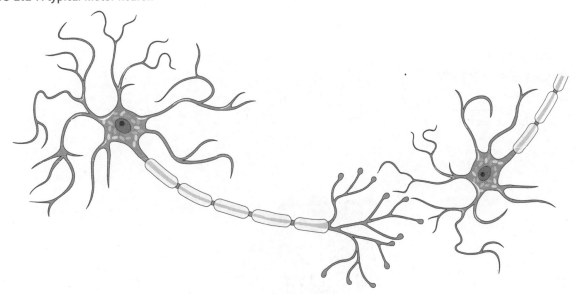

PART B

Table 2.6 lists functions that are performed by neurons.

1 Number each of the functions listed, in the order in which they occur.

Table 2.6 Neuron functions

Function	Order
The neurotransmitters received are excitatory or inhibitory. If excited, the nerve impulse (electrochemical energy) sweeps down the axon.	
Neurotransmitters attach to the receptor site on the dendrites of the postsynaptic neuron.	
The axon transmits information to the axon terminals.	
Neurotransmitters are released from the presynaptic neuron, across the synaptic cleft.	
Dendrites send information to the soma of the postsynaptic neuron.	
Neurotransmitters are returned to the presynaptic neuron and stored in vesicles.	

2.3.2 Neurotransmitters

Key science skills

Analyse, evaluate and communicate scientific ideas
- use appropriate psychological terminology, representations and conventions, including standard abbreviations, graphing conventions and units of measurement

Develop

This activity will strengthen your understanding of the role of neurotransmitters in synaptic transmission.

PART A

1 Use the following terms to label Figure 2.3, by writing the label and then drawing an arrow to indicate where each structural component is located.
- Synaptic cleft
- Synaptic vesicle
- Postsynaptic dendrite
- Neurotransmitter
- Presynaptic axon terminal
- Receptor site

Figure 2.3 Neural transmission

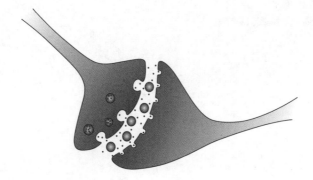

PART B

1 Using your textbook, other psychology textbooks and the Internet, research one of the following neurotransmitters involved in psychological functioning and behaviour: GABA, glutamate, dopamine or serotonin.

   a Neurotransmitter:
   _____

   b Main function(s):
   _____
   _____
   _____
   _____

   c Brain area(s) affected:
   _____
   _____
   _____
   _____

   d Effects on psychological functioning and behaviour:
   _____
   _____
   _____
   _____
   _____
   _____
   _____

2.3.3 Nervous system 'match the pairs'

> **Key science skills**
> Analyse, evaluate and communicate scientific ideas
> - discuss relevant psychological information, ideas, concepts, theories and models and the connections between them

Develop

This activity will revise the material on divisions of the nervous system and their specific structures and functions. It will also test your understanding of the role of neurotransmitters in the transmission of neural information.

1 Study the information pairs in Table 2.7.
2 Choose the correct information pairs to complete the statements. Write the words in the spaces provided.

Table 2.7 Information pairs

peripheral; brain	somatic; peripheral
inhibitory; excitatory	involuntary; visceral
autonomic; central	increase; decrease
somatic; autonomic	parasympathetic; sympathetic
glutamate; GABA	electrical; chemical
sympathetic; parasympathetic	dopamine; serotonin
sympathetic; flight-or-fight-or-freeze response	excitatory; inhibitory
voluntary; skeletal	autonomic; somatic
chemical; electrical	motor; sensory
voluntary; involuntary	neurotransmitters; neuromodulators

The order of the words in the information pairs is the same as the order in which they appear in the statements.

1. The major role of the _____ nervous system, as part of the _____ nervous system, is to carry motor messages to the skeletal muscles that control voluntary movement.
2. Information transmitted between neurons is sent in a _____ form. Once received, this information is transmitted along a neuron in an _____ form.
3. The _____ nervous system transmits sensory information inwards to the spinal cord and the spinal cord relays this information to the _____ for further processing.
4. Keeping the body in a state of physiological wellbeing is the responsibility of our _____ nervous system. However, when we become stressed or frightened our body needs extra energy to deal with the situation. The provision of this extra energy is the responsibility of our _____ nervous system.
5. The increased arousal we feel when we perceive a stressor automatically activates the _____ nervous system and we are instantly prepared for action. This nervous system then automatically triggers the _____ .
6. The _____ division of the autonomic nervous system energises our body by increasing heart rate, respiration, blood pressure, blood flow, release of sugar and secretion of hormones during times of high physiological arousal. In contrast, the _____ division of the autonomic nervous system decreases heart rate and blood pressure and stimulates digestion so the body is calmed and can return to its normal level of physiological functioning when the need for increased energy has passed.
7. The peripheral nervous system controls two kinds of neural pathways in the body: the _____, or 'voluntary' pathway, and the _____, or 'involuntary' pathway.
8. Changes in heart rate, blood pressure, digestion, muscle tension and perspiration are controlled by the _____ nervous system. Interpretation of sensory stimuli and the initiation of voluntary motor responses to stimuli are controlled by the _____ nervous system.
9. When you experience the flight-or-fight-or-freeze response, there is an _____ in the level of glucose released by your liver and a _____ in your digestion rate.
10. You are hot and perspiring, so you take off your jumper. Your _____ nervous system controls changes in your perspiration level, while your _____ nervous system is responsible for delivering the movement commands that enable you to remove your jumper.
11. The nerve impulse within a neuron is primarily _____; however, communication between neurons is _____ .
12. _____ neurons transmit commands sent from the brain to muscles, glands and organs that result in movement; _____ neurons receive and transmit information about specific forms of external energy and internal stimulation, which is analysed and then responded to by the brain.
13. The somatic nervous system controls the _____ movement of your _____ muscles.

CHAPTER 2 / Nervous system functioning

14 _____ are chemicals released into the synapse between neurons, to assist transmission of information between these neurons. _____ are chemicals that can modify the response of a neuron to a specific neurotransmitter.

15 The autonomic nervous system controls the _____ movement of your _____ muscles.

16 _____ is involved in high-speed neural transmission, while _____ is involved in slow neural transmission.

17 _____ makes you feel motivated and accomplished after completing an activity that activates the reward system in your brain. _____ boosts your mood and makes you feel happier.

18 GABA is the major _____ neurotransmitter, while glutamate is the major _____ neurotransmitter.

19 The somatic nervous system is the division of the peripheral nervous system that is responsible for controlling your _____ responses. The autonomic nervous system is the division of the peripheral nervous system that controls the body's _____ muscles, which control the activity level of internal organs and glands.

20 When a neuron increases its response to information received from another neuron, this activity is described as _____. When the receiving neuron reduces its response, this activity is described as _____.

2.4 Neural basis of learning and memory

Key knowledge
- synaptic plasticity – resulting from long-term potentiation and long-term depression, which together act to modify connections between neurons (sprouting, rerouting and pruning) – as the fundamental mechanism of memory formation that leads to learning

2.4.1 Neurological basis of memory and learning

Key science skills
Analyse, evaluate and communicate scientific ideas
- discuss relevant psychological information, ideas, concepts, theories and models and the connections between them

Develop

This activity will help you to understand the neurological basis of memory formation and learning and to understand the concepts of long-term potentiation (LTP) and long-term depression (LTD).

PART A

1 Table 2.8 contains statements about LTP and LTD. Place a tick in the correct box to indicate whether each statement is true or false.

Table 2.8 LTP and LTD

	True	False
Synaptic plasticity remains constant throughout life.		
Sprouting involves the growth of more branches of an axon so it can make more connections with other neurons.		
LTP is the result of the increased amount of adrenaline in the system when learning occurs.		
LTD is thought to improve the efficiency of the nervous system by reducing synaptic connections.		

	True	False
Rerouting is when a neural pathway creates an alternative path around a damaged area to connect with undamaged neurons.		
Pruning is when some of the axons of a neuron are removed to enable more efficient neuronal connections.		
Glutamate is a neurotransmitter involved in learning and memory.		
Learning is a relatively permanent change in behaviour due to experience.		
A neurotransmitter acts on adjacent cells by crossing the synaptic cleft and either inhibits or excites the adjacent cell.		
Synaptic plasticity includes the ability to 'prune' unused or excess connections.		
During LTP, changes occur in the post-synaptic neuron.		
Glutamate is not involved in LTP nor in LTD.		
LTP results in the creation of more glutamate receptors on the post-synaptic neuron.		
Neural plasticity involves the neuron changing in structure.		

Exam practice

Multiple choice

Circle the response that best answers the question.

1. Which of the following functions is controlled by the autonomic nervous system?
   A voluntary movement of the skeletal muscles and perspiration rate
   B perspiration rate, heart rate and blinking
   C voluntary movement of the skeletal muscles and liver function
   D voluntary movement of the skeletal muscles and saliva production

   Use the following information to answer Questions 2–5.
   Isaac is completing his homework. He reads each question, thinks about a response and then writes his answer in his workbook. As he is completing his homework, a large black spider crawls along his desk. Instantly he feels an increase in his heart and breathing rates.

2. Which part of Isaac's nervous system is mostly responsible for thinking about and deciding on a response for each homework question?
   A the somatic nervous system
   B the autonomic nervous system
   C the central nervous system
   D the sympathetic nervous system

3. The ability for Isaac to move his hand to write his responses is due to the action of
   A the somatic nervous system.
   B the parasympathetic nervous system.
   C the central nervous system.
   D the sympathetic nervous system.

4. When Isaac noticed the spider, the nervous system responsible for the increase in his heart and breathing rates was the
   A somatic nervous system.
   B parasympathetic nervous system.
   C central nervous system.
   D sympathetic nervous system.

5. A physiological response that Isaac was not likely to have experienced when he noticed the spider was
   A dilation of pupils.
   B increase in hormone levels.
   C increase in the production of saliva.
   D increased perspiration.

6. Imogen was cooking when her hand came in contact with the stovetop. She automatically withdrew her hand from the surface. This action is the result of
   A the activity of the somatic nervous system.
   B the activity of the autonomic nervous system.
   C an action potential.
   D a spinal reflex.

7. Which of the following statements about the autonomic nervous system is incorrect?
   A It is mostly self-regulating, operating independently of the brain.
   B Techniques such as yoga and meditation can enable us to gain control over our autonomic nervous system.
   C It is a division of the peripheral nervous system that controls the activity level of our internal organs and glands.
   D It only operates during times of high emotional or physical arousal.

8 Which statement best describes the structure of the spinal cord?
   A The centre of the spinal cord is made up of white matter and the outer layers are made up of grey matter.
   B Grey matter is made of cell bodies, their axons and their dendrites and is found in the centre of the spinal cord. The outer layers of the spinal cord are made up of columns of white matter containing myelin-coated axons.
   C White matter is made of cell bodies, their axons and their dendrites and is found in the centre of the spinal cord. The outer layers of the spinal cord are made up of columns of grey matter containing myelin-coated axons.
   D Grey matter is made of myelin-coated axons only and forms ascending and descending tracts in the outer layers of the spinal cord. The inner layers of the spinal cord are made up of columns of white matter containing cell bodies, their axons and their dendrites.

9 Which of the following statements about neurotransmitters is incorrect?
   A Neurotransmitters enable communication between neurons around the nervous system.
   B Neurotransmitters are stored in synaptic vesicles and are released into the synaptic cleft when the message reaches the soma of the presynaptic neuron.
   C Neurotransmitters carry the chemical message across a synapse to the receptor site of the postsynaptic neuron.
   D Neurotransmitters attach themselves to specific receptor sites in a way that resembles a key fitting a specific lock.

10 Neurotransmitters can act in one of two ways when they arrive at the postsynaptic neuron. If the neurotransmitter has an _____ effect, it is more likely that the neuron will fire an action potential. If the neurotransmitter has an _____ effect, it is less likely that the neuron will fire an action potential.
   A agonist; antagonist
   B antagonist; agonist
   C excitatory; inhibitory
   D inhibitory; excitatory

11 Like neurotransmitters, drugs have their own molecular shape and effects. A drug that is considered an _____ will either increase the release of neurotransmitters or imitate certain neurotransmitters, while a drug classified as an _____ will either inhibit the release of neurotransmitters or block the receptor sites for these neurotransmitters.
   A agonist; antagonist
   B antagonist; agonist
   C excitatory; inhibitory
   D inhibitory; excitatory

12 Which statement about the neurotransmitter GABA is incorrect?
   A GABA is a type of amino acid found extensively throughout the nervous system.
   B GABA is one of the neurotransmitters influential in the development and treatment of anxiety disorder.
   C GABA is the major inhibitory neurotransmitter in the nervous system.
   D When GABA activates its receptor sites, the cells that have those receptors are excited.

13 Which statement about the neurotransmitter glutamate is incorrect?
   A Caffeine is thought to increase glutamate activity.
   B Glutamate is one of the neurotransmitters influential in the development and treatment of anxiety disorder.
   C Glutamate is the major excitatory neurotransmitter in the nervous system.
   D Glutamate has a vital role in cognitive functions and LTP.

14 The term used to describe the brain's ability to change and adapt as conditions change is
   A synapse.
   B plasticity.
   C genetics.
   D neurosis.

15 Which of the following is thought to occur during most learning experiences?
   A An increase in synaptic plasticity and a decrease in neural plasticity.
   B A release of adrenaline, increasing the likelihood of learning being consolidated to memory.
   C A release of glutamate leading to LTP and excitation of neurons.
   D A release of glutamate leading to LTD and excitation of neurons.

Short answer

1 Use Figure 2.4 to complete the following tasks.
   a Complete the diagram to show the divisions of the nervous system. 6 marks
   b Draw a circle around the box that contains the part of the nervous system that controls reflex actions. 1 mark

Figure 2.4

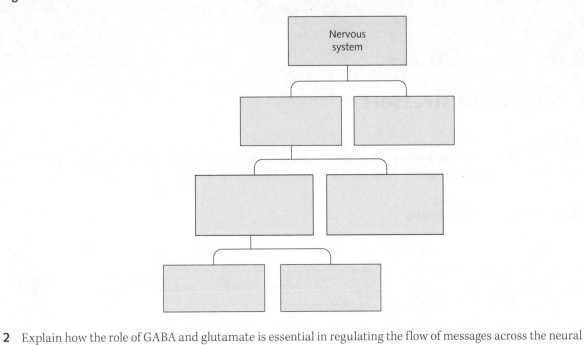

2 Explain how the role of GABA and glutamate is essential in regulating the flow of messages across the neural synapse. 3 marks

_____
_____
_____
_____
_____

3 Use the space below to create a table to show the similarities and differences between LTP and LTD. 6 marks

3 Stress as an example of a psychobiological process

3.1 Stress and stressors

Key knowledge
- internal and external stressors causing psychological and physiological stress responses, including the fight-or-flight-or-freeze response in acute stress and the role of cortisol in chronic stress

3.1.1 Introduction to stress

Key science skills
Analyse, evaluate and communicate scientific ideas
- use appropriate psychological terminology, representations and conventions, including standard abbreviations, graphing conventions and units of measurement
- discuss relevant psychological information, ideas, concepts, theories and models and the connections between them

To understand the material presented in this chapter you have to make sure you understand the key terms used. This activity will help you to clarify your understanding of the relationship between a stressor, stress and a stress response.

PART A

1. In Figure 3.1, define the terms 'stressor', 'stress' and 'stress response' in the spaces (ovals) provided.
2. Define the terms 'acute stress', 'chronic stress', 'eustress' and 'distress' in the spaces (rectangles) provided.
3. Outline the psychological effects of eustress and distress, and the physiological effects of the stress response.

Figure 3.1 The relationship between a stressor, stress and the stress response

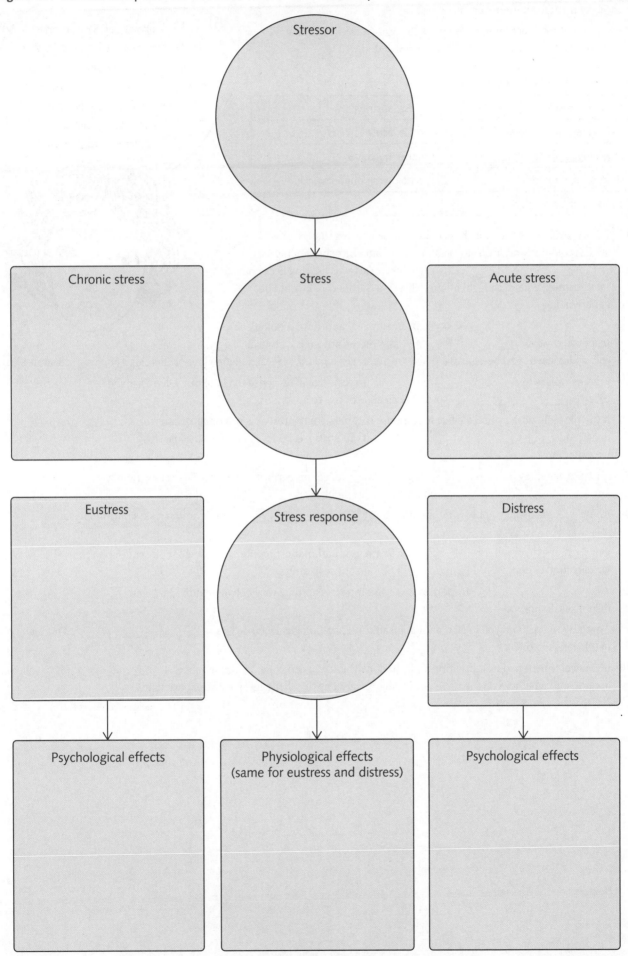

PART B

1. Study the information pairs in Table 3.1.
2. Choose the correct information pairs to complete the statements below. Write the words in the spaces provided.

Table 3.1 Information pairs

stressor; stress	distress; eustress
a stress response; stress	chronic; acute
physiological; psychological	positive; negative
short term; long term	psychological; physiological
eustress; distress	

The order of the words in the information pairs is the same as the order in which they appear in the statements. Some pairs may be used more than once.

1. If I was excited by the thought of delivering a speech at my sister's wedding and my heart beat faster, I would be experiencing _____. However, if my heart was beating faster because I felt terrified when I thought about delivering the speech, I would be experiencing _____.

2. _____ describes any physical and/or psychological response to stress. _____ is the automatic arousal (physiological or psychological) a person experiences when they feel challenged by, or unable to cope with, a situation.

3. Eustress refers to a _____ psychological response to a stressor. Distress refers to a _____ psychological response to a stressor.

4. If the physiological arousal we feel in response to a persistent stressor is prolonged, then we experience _____ stress. If our stress response is brief but intense, we experience _____ stress.

5. Eustress generally has a _____ effect on performance level. Distress generally has a _____ effect on performance level.

6. Acute stress is usually not harmful because it is _____. Chronic stress is harmful because it is _____.

7. Our _____ response to stress is automatic; therefore, it is beyond our conscious control. However, we can control our _____ response to stress.

8. _____ responses to a common stressor may vary between individuals but _____ responses do not.

9. Anything that causes you to feel physiologically and psychologically tense is a _____. This feeling of tension is known as _____.

10. If you feel threatened by a situation and you don't think you can cope, you experience _____. If you are not sure whether you can cope but you feel excited by the challenge, you are experiencing _____.

CHAPTER 3 / Stress as an example of a psychobiological process   45

3.1.2 Sources of stress

Key science skills
Analyse, evaluate and communicate scientific ideas
- discuss relevant psychological information, ideas, concepts, theories and models and the connections between them

This activity will review and strengthen your understanding of the hierarchical structure of the human nervous system. You could challenge yourself to complete this task without referring to your notes. How much can you do?

This activity will also improve your understanding of how stress can result from internal stressors or external stressors including daily pressures, significant life events or major catastrophes.

PART A

Study Figure 3.2.
1. Define 'internal stressors', 'daily pressures', 'life events' and 'major catastrophes' (as they apply to stress) in the spaces provided.
2. Identify three examples of each source of stress.
3. Identify three possible effects of each source of stress.

Daily pressures, life events and major catastrophes are all examples of external stressors.

Figure 3.2 Sources of stress

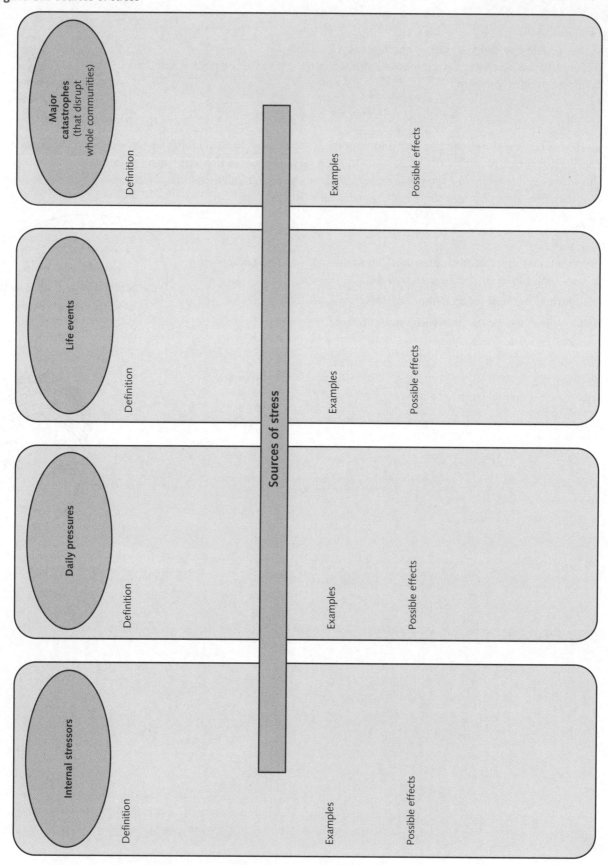

PART B

1 To check that you understand the different sources of stress, place a tick in the correct column of Table 3.2 to indicate which source of stress is represented in each statement.

Table 3.2 Examples of sources of stress

Statement	Internal stressor	Daily pressure	Life event	Major catastrophe
The state of Victoria being put into lockdown due to COVID-19				
Loss of family and social support as a result of migrating to Australia				
Moving out of home to begin university				
Being caught in the same traffic jam every day while trying to get to school				
Losing your mobile phone				
Loss of home and possessions in a bushfire				
Relocating to China and not knowing the language or the culture				
The birth of a sibling, which means you don't have your own room anymore				
Breaking your iPad screen				
Breaking your arm by falling down some stairs				

3.1.3 Flight-or-fight-or-freeze response

Key science skills
Analyse, evaluate and communicate scientific ideas
- discuss relevant psychological information, ideas, concepts, theories and models and the connections between them

Develop

This activity has been designed to strengthen your understanding of the body's flight-or-fight-or-freeze response to stress.

PART A

1 Read the following list.
- Flight-or-fight-or-freeze response blocked
- Sympathetic nervous system automatically activated
- Blood diverted to skeletal muscles
- Pupils contract
- Energises and arouses the body so it is prepared for action
- Parasympathetic nervous system automatically suppressed
- Sympathetic nervous system automatically suppressed
- Pupils dilate
- Blood diverted to body's core

The correct number of bullet points have been placed for you in the appropriate sections.

- Parasympathetic nervous system automatically activated
- Muscles relax
- Immobilises the body and conserves energy
- Muscles tense

2 Write each bullet point in the correct space in Figure 3.3.

Figure 3.3 The flight-or-fight-or-freeze response

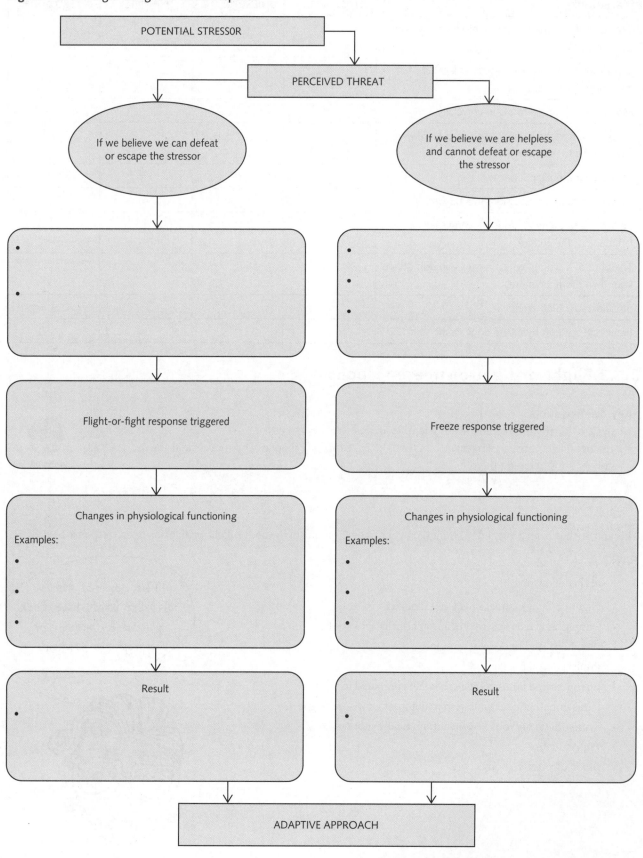

PART B

Use the terms in Table 3.3 to answer the following 'What am I?' questions.

Table 3.3 Information related to the flight-or-fight-or-freeze response options to stress

Sympathetic nervous system	Tonic immobility	Fight response
Walter Bradford Cannon	Freeze response	Parasympathetic nervous system
Flight response	Hans Selye	Relaxation response

Not all the terms will be used and some terms will be used more than once.

1   I am the response to a perceived threat that results from a sense of helplessness in the face of danger. What am I?

2   I am the division of the nervous system that is activated if you feel you can take some action to deal with a stressor. What am I?

3   I am the division of the nervous system that is suppressed if you feel powerless when confronted by a stressor. What am I?

4   I am the response triggered by the parasympathetic nervous system when danger is thought to have passed. What am I?

5   I am the response to a perceived threat when you believe that you have the ability to confront the threat. What am I?

6   In the 1920s, I introduced the concept of the flight-or-fight-or-freeze response to a perceived threat. Who am I?

7   When you are faced with a threat, I am the structure that helps you to survive by conserving your energy and directing it only to the organs in your body's core that you need for basic survival until the danger has passed. What am I?

8   When you think you can do nothing to survive an extreme threat, I am triggered, as the activation of the parasympathetic nervous system is so strong that it suppresses the activation of the sympathetic nervous system. What am I?

9   I am the response to a perceived threat when you believe that you have the ability and opportunity to escape from a threat. What am I?

10   I am responsible for maintaining homeostasis. What am I?

3.1.4 Effects of stress

Key science skills
Analyse, evaluate and communicate scientific ideas
- discuss relevant psychological information, ideas, concepts, theories and models and the connections between them

Develop

This activity will help you to clarify your understanding of the physiological and psychological effects of prolonged stress.

PART A

1   Place a tick in the correct column of Table 3.4 to indicate whether each statement is true or false.

Table 3.4 Effects of stress

Statement about stress	True	False
Stress increases susceptibility to disease – it does not cause disease.		
Stress, in the form of distress, can help the body perform at optimal levels.		
A stress reaction begins with the perception of a stressor.		
Prolonged stress results in the depletion of all hormones.		
A short-term increase in levels of cortisol allows the body to cope with a stressful event.		
The effects of adrenaline last a lot longer than the effects of cortisol.		
A stress response occurs when the flight-or-fight-or-freeze response triggers the sympathetic nervous system.		
Cortisol, adrenaline and noradrenaline are released in the same way at the same time.		
If my stress is intense and uncontrollable and hinders my ability to perform at my best level, I will be experiencing eustress.		
Cortisol is only produced by the body in response to stress.		

PART B

Table 3.5 contains examples of the effects of stress. Write the symptoms in the correct columns of Table 3.6 to indicate whether they are physiological or psychological symptoms.

Table 3.5 Symptoms of stress

Difficulty swallowing	Angry outbursts	Sleep disturbances	Diarrhoea
Feeling frustrated	Dizziness	Lack of concentration	Headache
Fidgeting	Jumping at sudden noises	Respiratory disorders (e.g. asthma)	Skin rashes (e.g. eczema)
Muscle tension	Jaw clenching or teeth grinding	Fatigue	Irritability
Believing you are unable to cope	Muscular aches and pains	Heart palpitations	Negative thoughts and feelings

Table 3.6 Physiological and psychological symptoms of stress

Physiological	Psychological

CHAPTER 3 / Stress as an example of a psychobiological process

3.2 Selye's General Adaptation Syndrome

Key knowledge
- the explanatory power of Hans Selye's General Adaptation Syndrome as a biological model of stress, including alarm reaction (shock/counter shock), resistance and exhaustion

3.2.1 The General Adaptation Syndrome

Key science skills
Analyse, evaluate and communicate scientific ideas
- discuss relevant psychological information, ideas, concepts, theories and models and the connections between them

Develop

This activity will reinforce your understanding of the relationship between stress and disease, and the General Adaptation Syndrome (GAS).

PART A

Figure 3.4 is a model of Hans Selye's GAS. Complete this model by following the instructions below.

1. Write the names of the stages of the GAS in the spaces provided.
2. Draw a curve in the top part of the figure to show a person's resistance level in each stage of the GAS, when compared to the normal level of resistance to stress.
3. Describe what happens to a person in each stage of the GAS by writing the following bullet points in the columns under the appropriate stage.
   - If the stressor remains, the body's level of resistance is higher than normal for the stressor but lower than normal for additional stressors.
   - The stressor is perceived.
   - Resistance to stress drops below normal.
   - The sympathetic nervous system is activated and the flight-or-fight-or-freeze response is triggered.
   - The body acts as if it is injured.
   - Stress hormones are released into the bloodstream, causing physiological changes that increase the resistance level to above normal.
   - Cortisol is released to help repair the damage caused by stress on the body.
   - The immune system starts to become compromised and signs of illness appear.
   - Bodily resources, including hormones, are drained and resistance falls below normal.
   - The immune system can no longer function effectively and the body succumbs to illness or injury.

Figure 3.4 A model of Hans Selye's General Adaptation Syndrome

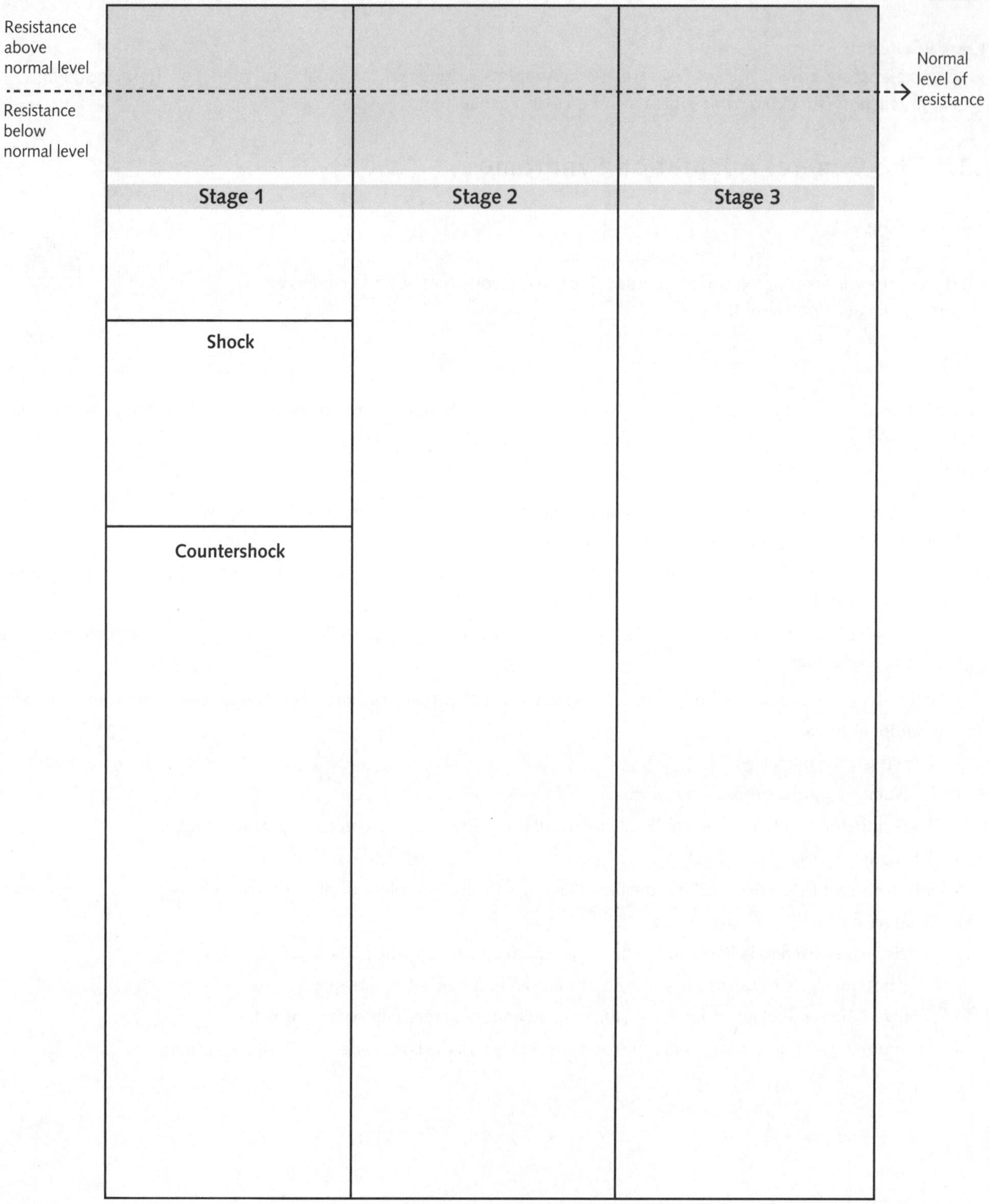

	Stage 1	Stage 2	Stage 3

Resistance above normal level
- → Normal level of resistance
Resistance below normal level

Stage 1: Shock / Countershock

PART B

Read each scenario and then answer the questions that follow.

1. Oscar was reclining on the couch watching television when he suddenly noticed a large, dark shape moving across the floor towards him. Oscar assumed it was a spider and immediately leapt off the couch and ran out of the room.

   a. Name the response that was activated when Oscar noticed the spider.

b Explain why Oscar was able to move so quickly.

_____
_____
_____
_____

c When Oscar returned to examine the shape, he realised it was only a dried leaf, blown in by the breeze from an open window. He calmed down. Which branch of Oscar's nervous system was responsible for this reaction?

_____

2 Two months after starting a very demanding job, Sunita was busily preparing for her 21st birthday party. Although she was feeling tired and beginning to have headaches, she was enjoying the preparations. Three days before the party, she received news that her much loved uncle had suffered a heart attack. On the day of her party, she developed a migraine, couldn't stop crying and was feeling so ill she couldn't get out of bed.

a Which stage of the GAS was Sunita experiencing when she began to have headaches?

_____

b Which stage of the GAS was Sunita experiencing on the day of her party?

_____

c According to the GAS, why did Sunita fall ill on the day of her party?

_____
_____
_____
_____
_____

3 Summarise the relationship between stress and illness.

_____
_____
_____
_____
_____

3.2.2 The GAS: a case study

This activity will strengthen your understanding of how the body reacts to stress according to the theory of the GAS. Read the case study and answer the questions that follow.

CASE STUDY

JULIA

Six months after moving out of the family home, 20-year-old Julia was enjoying her independent lifestyle in a new two-bedroom city apartment. Although she was on a very tight budget and could only spend money on necessary items, Julia did not regret her decision to live alone. Julia kept her apartment very neat and clean, and she enjoyed the peace and quiet. Julia was very health conscious, and she particularly enjoyed being able to unwind from work by doing yoga in her second bedroom, which she had set up as an exercise room.

One day Julia received a text message from her only cousin, Cindy, saying that she and her five-year-old son, Jason, were moving to the city. Cindy asked if they could stay with Julia while they set themselves up in their own apartment. Julia agreed and the next night she received another text message saying that Cindy and Jason would be arriving at the end of the week. The thought of them arriving so soon upset Julia. She liked her cousin, but when Cindy had visited Julia's family in the past, she had been very noisy and messy and Jason was badly behaved. Additionally, Julia would have to rearrange her finances to purchase two beds for the spare room, and move her exercise equipment out of that room. However, Julia knew they had nowhere else to stay. Julia woke up regularly during the following nights, worrying and making arrangements in her head. She forgot to eat during the day as she was absorbed in writing lists and making calls. She started to wonder if she would be ready in such a short time.

A week after Cindy and Jason arrived, Julia's apartment was a mess. Cindy said she was too busy entertaining Jason to wash dishes or do chores, and Jason left toys and rubbish throughout the apartment. Jason would also throw tantrums when he didn't get what he wanted and would only eat his meals if he could sit on the floor in front of the television. Julia was upset by all of this and didn't know what to do. She tried talking to Cindy about sharing the chores and she tried to keep Jason calm but the harder she tried, the messier and noisier her apartment seemed to become. Julia tried to ignore the mess and the noise but she could not relax when she was at home. She also started to get headaches.

After three months of living with Cindy and Jason, Julia decided the only way to have a clean and tidy apartment was if she made twice the effort to clean it. However, every time she did this, the mess returned within 48 hours and the noise never seemed to cease. As Julia became more upset by these circumstances, the difficulty sleeping returned. She started to stay at work longer in the hope that she would be very tired when she went home. However, she found that instead of sleeping more, she was having even more difficulty sleeping and was feeling very tired during the day. Julia also noticed that at work she was forgetting some of the tasks she was given.

This situation continued for 12 months and one day when Julia was at work Cindy phoned her to say that Jason had set fire to the rug in the lounge room but the fire was put out before it spread to the rest of the apartment. Julia felt sick when she heard this because, after paying her mortgage every month, she didn't have any savings and she hadn't insured the contents of her apartment.

In the weeks that followed, Julia's sleeping problems and daytime tiredness increased. She felt physically and emotionally drained. She had no appetite and her face broke out in pimples. One day at work Julia started to have difficulty breathing and had to be taken to hospital. The doctor diagnosed Julia as suffering from a respiratory infection.

1 Name the researcher who created the model known as the GAS.

2 Prior to receiving the first text message from Cindy, was the GAS model relevant to Julia? Explain your answer.

3 Name the stage of the GAS Julia experienced when she received the second text message from Cindy.

4 Identify three physiological changes that Julia would have experienced during the countershock phase of Stage 1 of the GAS.

5 What biological response triggered Julia's physiological changes?

6 Identify the signs that indicate Julia had entered the stage of exhaustion.

7 In reference to the effects of stress on Julia's body, explain why she entered the stage of exhaustion.

3.3 Stress as a psychological process

Key knowledge
- the explanatory power of Richard Lazarus and Susan Folkman's Transactional Model of Stress and Coping to explain stress as a psychological process (primary and secondary appraisal only)

3.3.1 Lazarus and Folkman's Transactional Model of Stress and Coping

Key science skills
Analyse, evaluate and communicate scientific ideas
- discuss relevant psychological information, ideas, concepts, theories and models and the connections between them

In 1984, Richard Lazarus and Susan Folkman introduced a model that explained the mental process, or cognitive appraisal, which influences our response to stressors. This model is referred to as the Transactional Model of Stress and Coping. This activity will reinforce your understanding of Lazarus and Folkman's Transactional Model of Stress and Coping.

PART A

Figure 3.5 provides you with an outline of Lazarus and Folkman's Transactional Model of Stress and Coping.
1 Complete this model by:
   a defining the Transactional Model of Stress and Coping in the space provided
   b defining 'primary appraisal' and 'secondary appraisal' in the spaces provided
   c explaining distress and eustress within the primary and secondary appraisals, and then writing examples in the spaces provided.

Figure 3.5 Lazarus and Folkman's Transactional Model of Stress and Coping

- TRANSACTIONAL MODEL OF STRESS AND COPING
  - PERCEPTION OF POSSIBLE STRESSOR
    - Example: Being told that your workplace is closing down
  - PRIMARY APPRAISAL
    - DISTRESS is experienced if you...
      - Example
    - EUSTRESS is experienced if you...
      - Example
  - SECONDARY APPRAISAL
    - DISTRESS is experienced if you...
      - Example
    - EUSTRESS is experienced if you...
      - Example

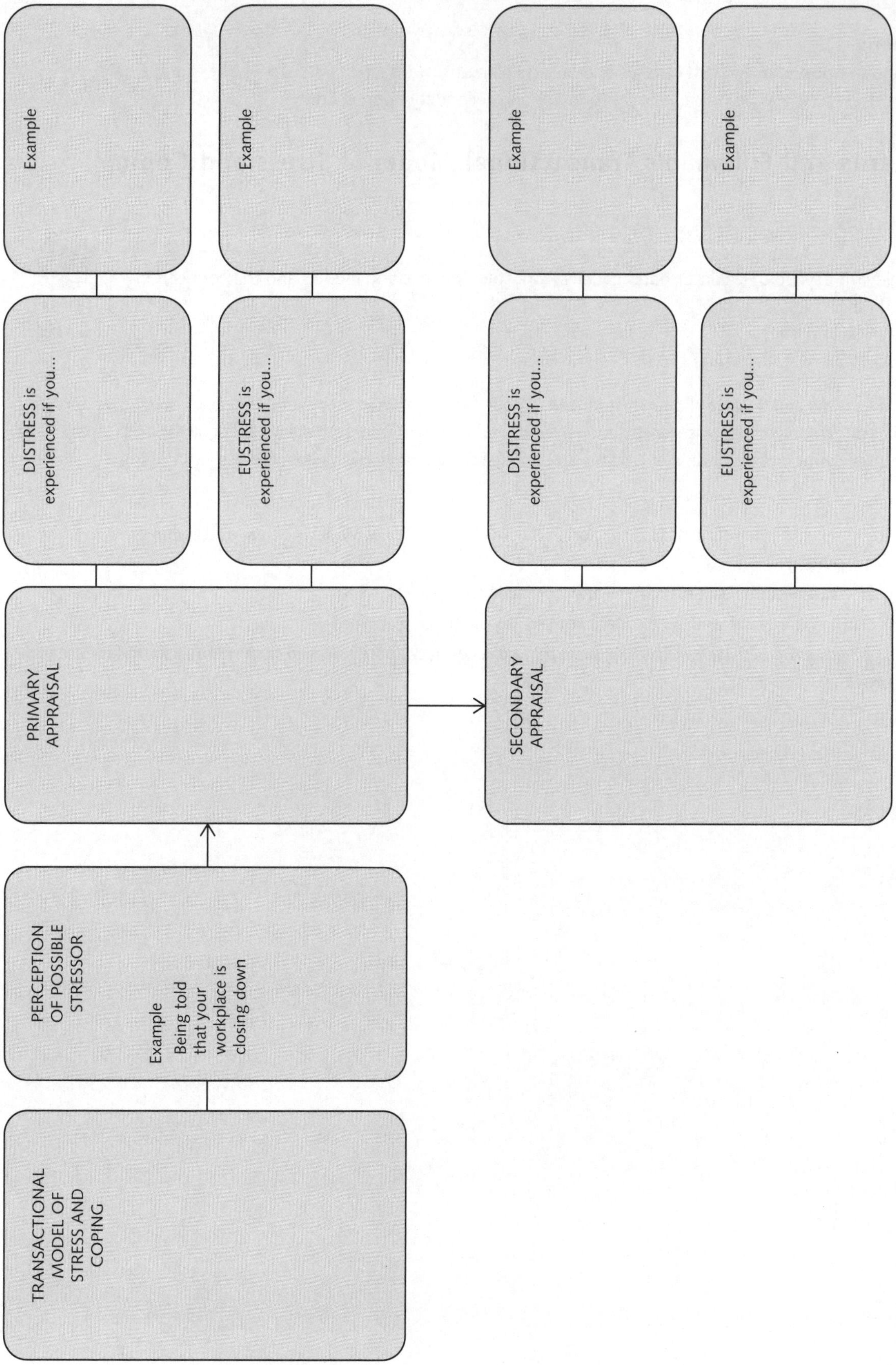

PART B

1 Use the terms in Table 3.7 to fill in the blanks in the paragraphs that follow.

Table 3.7 Terms associated with Lazarus and Folkman's Transactional Model of Stress and Coping

transaction	external environment	primary	cognitive appraisal
sequential	secondary	resources	physiological
reducing	negative	subjective	relevant
psychological	control	threatening	reappraise

While Selye's GAS focuses on the _____ determinants of stress, Lazarus and Folkman's Transactional Model of Stress and Coping is a _____ approach to stress because it outlines the mental processes that influence our response to stressors. It suggests that stress is a _____ experience that involves an encounter, or _____, between an individual and their _____. This model suggests that a stress response depends on the individual's evaluation, or _____, of the stressor and their perceived ability to cope with it. According to this model, we experience stress when we are faced with situations we feel unable to cope with. The Transactional Model of Stress and Coping encompasses two _____ stages of cognitive appraisal.

Lazarus and Folkman suggest that initially we engage in _____ appraisal because we decide whether a situation is _____ or irrelevant to us, benign–positive (harmless–desirable) or potentially _____ and harmful. If we decide that the potential stressor can do us no future harm because it is irrelevant, benign–positive or non-threatening, we do not experience stress. If we appraise it as a threat, we view it as something that can cause future harm and we move to a stage of _____ appraisal, the stage when we assess whether or not we have the internal and/or external _____ to cope with the stressor and _____ our situation. If we feel we have the resources to cope, we may _____ the stressor as being less significant and less threatening, thereby _____ our stress. If we feel we cannot cope with the threat, we experience a _____ stress reaction.

3.4 The gut–brain axis

Key knowledge
• the gut–brain axis (GBA) as an area of emerging research, with reference to the interaction of gut microbiota with stress and the nervous system in the control of psychological processes and behaviour

3.4.1 The gut–brain axis (GBA)

Key science skills
Analyse, evaluate and communicate scientific ideas
• discuss relevant psychological information, ideas, concepts, theories and models and the connections between them

Develop

The GBA is the network of bidirectional (two-way) neural pathways that enable communication between bacteria in the gastrointestinal (GI) tract and the brain. This activity will help you to understand the structure of the GBA and the role that it plays in influencing our mental wellbeing and behaviour.

PART A

1 Place the following terms on the correct side of Figure 3.6, depending on whether the term affects the brain, or the gut.
   • decreases digestion when in flight-or-fight-or-freeze mode
   • maintains digestive acid
   • absorbs tryptophan, a precursor of serotonin

- influences stress/anxiety
- controls secretion of mucous
- indicates level of fullness in the stomach
- triggers mood changes and feelings of depression
- influences bowel movement frequency and solidness
- increases or decreases digestion
- relays information about bacteria in the gut

2 Draw an arrow on Figure 3.6 to show the connection of the brain and the gut via the vagus nerve. Think whether your arrow should be unidirectional or bidirectional.

Figure 3.6 The gut–brain axis

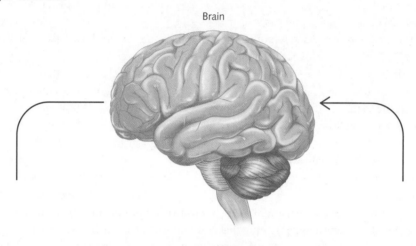
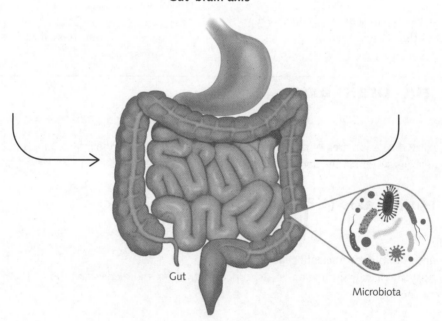

CHAPTER 3 / Stress as an example of a psychobiological process

PART B

1 Complete the summary of the gut–brain axis using the terms provided in Table 3.8.

Table 3.8 Terms related to the gut–brain axis

gastrointestinal	saliva	microbiome	bidirectional
immune system	lungs	dysbiosis	enteric
microbiota	central	stress	vagus
intestines	brain	autonomic nervous system	motor

Hint: Cross off each term as you use it.

The GBA is the network of _____ (two-way) neural pathways that enable communication between bacteria in the gastrointestinal (GI) tract and the brain. The GBA includes the central nervous system, the autonomic nervous system, the _____ nervous system, the _____ nerve and the gut _____. It is also linked to the immune and endocrine (hormonal) systems. The enteric nervous system is a branch of the _____, which functions independently of the _____ nervous system to manage the functions of the digestive system. Communication between the gut and the brain happens primarily through the vagus nerve. The vagus nerve is the longest nerve in the human body, running from the brain stem all the way down to the _____. It connects many abdominal organs including the gut, heart, _____ and liver. The vagus nerve connects with most of the GI system and is the major communication route between the gut and the _____. Ninety per cent of the vagus nerve fibres are afferent (sensory) connections that send signals 'up' from the gut to the brain. The remaining 10 per cent of nerve fibres send efferent _____ signals from the brain to the gut, such as the release of _____ and stomach acids, information about incoming food and required changes in movements to aid digestion. The gut microbiota (also referred to as _____) is the highly diverse and dynamic system of almost 100 trillion bacteria and other micro-organisms that live in the human _____ tract. _____ (when the gut bacteria become less diverse or there is no longer a healthy balance of bacteria) can cause a range of digestive illnesses and reduce the effectiveness of the _____ overall. Thus, any disruption in the health of the microbiome can increase the body's susceptibility to disease. Through this research, scientists have found that the health of the microbiota can influence our social behaviour, cognition and feelings of depression, anxiety and _____.

3.4.2 The human microbiome project

Key science skills
Analyse, evaluate and communicate scientific ideas
- discuss relevant psychological information, ideas, concepts, theories and models and the connections between them

Develop

The gut microbiota or microbiome is the highly diverse and dynamic system of almost 100 trillion bacteria and other micro-organisms that live in the human GI tract. This activity will help you understand the importance of the microbiome.
Read the following information about the human microbiome project (HMP), then answer the questions that follow.

Human microbiome project

The HMP began in 2007 and took until 2016 to complete. It was coordinated and funded by the National Institutes of Health (NIH) in the USA. The research focused on identifying the microbial communities that live in and on our bodies and how they influence human metabolism. To do this, the DNA of each microbe that was found in healthy adults was sequenced. Samples from over 300 healthy humans were taken and a comprehensive profile of their microbiome was created (see Figure 3.7).

Figure 3.7 Sites on the body that were sampled for microbes in the HMP

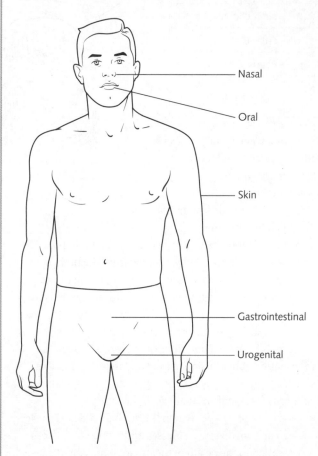

These profiles produced a baseline set of data. The microbes found in individuals with specific diseases (for example, irritable bowel syndrome) were then sequenced and compared to the data from healthy individuals. The data also recorded the small molecules and proteins that were being produced by the unique microbes in individuals that had a microbe-related disease.

Findings from the HMP include:
- Microbial cells outnumber human cells by about ten to one.
- ~3000 bacterial genomes were sequenced.
- The composition of microbes in individuals with inflammatory bowel disease fluctuated over time a lot more than in healthy individuals.
- In patients with Crohn's disease, gut dysbiosis increased when individuals were taking antibiotics, but anti-inflammatories reduced gut dysbiosis. Monitoring changes in the gut microbiome is now used as an early detection of Crohn's disease.
- A bacterial species named *Christensenella minuta* tends to be absent in people that are overweight.
- A short chain fatty acid called butyrate is produced by the microbiome. Butyrate has been found to influence the circadian rhythm of the individual. A link was found between a high-fat diet and the production of butyrate, which in turn altered circadian rhythms.
- The gut microbiome produces a compound, TMAO, when breaking down red meat. TMAO is known to have a role in promoting cardiovascular disease.
- The study concluded that changes in the human microbiome can be linked with changes in health and the emergence of disease. The information gained from the HMP is available to anyone across the world who might be pursuing research into the causes of human disease.

Source: Human Microbiome Project's Health Relevance, National Institutes of Health, U.S. Department of Health & Human Services

1 What was the aim of the HMP?

2 How long did it take to complete this research?

3 Why were healthy individuals used in this study?

4 What is gut dysbiosis?

5   Suggest some future research that may occur as a result of the finding that overweight individuals are missing the bacterium *Christensenella minuta*.

    _____
    _____
    _____

6   To fund the HMP would have been hugely expensive. Why has the NIH made the findings from this research freely available across the world? Why not try and recoup some of their costs?

    _____
    _____
    _____

3.5 Coping strategies

Key knowledge
- use of strategies (approach and avoidance) for coping with stress and improving mental wellbeing, including context-specific effectiveness and coping flexibility

3.5.1 Strategies for coping with stress

Key science skills
Analyse, evaluate and communicate scientific ideas
- discuss relevant psychological information, ideas, concepts, theories and models and the connections between them

Coping skills consist of behaviours or techniques that help us solve problems or meet demands. Coping is a complex process that varies according to the demands of the stressor, the individual's appraisal of the situation and the personal and social resources available. This activity will help you strengthen your understanding of the strengths and limitations of different strategies for coping with stress.

PART A

Figure 3.8 provides an outline of the strategies for coping with stress. Complete Figure 3.8 by following the instructions below.

1   Write a definition for each of the following terms in the boxes provided.
    - coping strategies
    - physical exercise
    - approach strategies
    - problem-focused strategies
    - emotion-focused strategies
    - avoidance strategies
2   List three physical benefits and three psychological benefits of physical exercise in the boxes provided.
3   List three benefits of approach strategies in the box provided.
4   List three limitations of avoidance strategies in the box provided.

Figure 3.8 Strategies for coping with stress

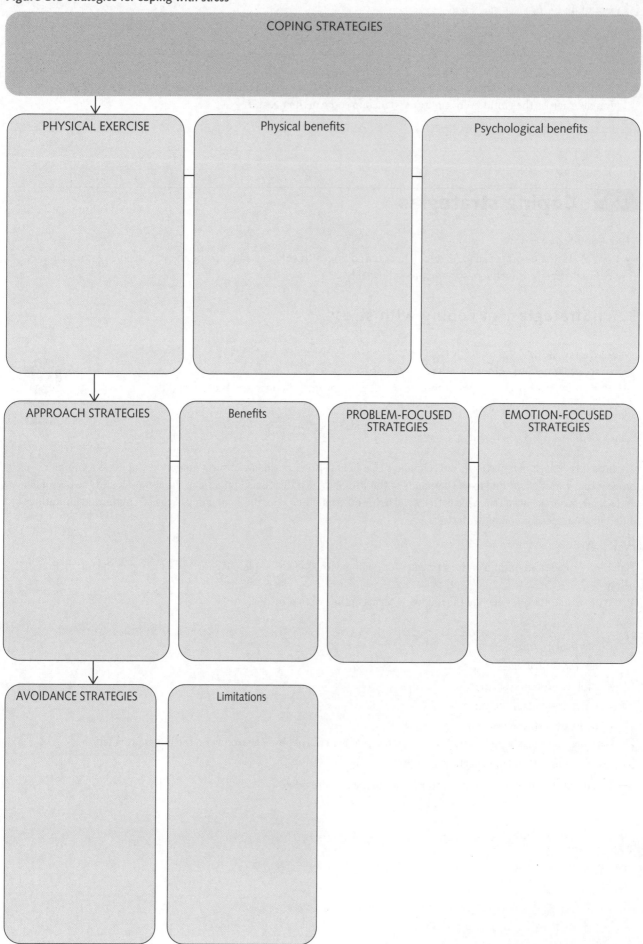

PART B

Use the appropriate pair of terms in Table 3.9 to fill in the blanks in the following statements.

Table 3.9 Terms related to stress-management strategies

problem-focused; emotion-focused	approach; avoidance	adaptive; maladaptive
venting; denial	endorphins; cortisol	denial; venting
coping; coping flexibility	cortisol; endorphins	

Hint: The order of the words in the pairs is the same as the order they appear in the statements.

1. _____ is a form of avoidance strategy; _____ is a form of approach strategy.

2. Physical exercise helps our body to cope with stress because it promotes the release of _____, which lift our mood and provide physical pain relief. In addition, it helps our body to absorb excess _____, which could damage us in the long term.

3. _____ is an example of an energising chemical released by the adrenal glands; _____ are an example of a mood-lifting chemical released by the brain.

4. _____ strategies help us cope with the stressor in practical ways; therefore, they are considered to be positive and adaptive. However, _____ strategies are considered to be negative and maladaptive because they often increase our stress level.

5. If your coping strategy is aimed at reducing the impact of the stressor, it is _____. If it is aimed at changing the way you feel about it, it is _____.

6. Making a plan of action to cope with a stressor is an example of an _____ coping strategy because it could boost your confidence in your ability to cope. Binge eating to make yourself feel better is an example of a _____ coping strategy because it could prove harmful to you.

7. Emotion-focused strategies can be both positive and negative. _____ is a positive emotion-focused strategy because it releases negative emotion and relaxes you so you can think clearly about how to deal with the stressor. _____ is a negative emotion-focused strategy because, although it may make you feel better, it does not help you find ways to deal with the stressor.

8. The process of constantly changing thoughts and behaviours so we can manage the demands of stressors is referred to as _____. A person's ability to adapt their behaviours to suit the demands of the stressor is referred to as _____.

3.5.2 Evaluation of research

Key science skills

Develop aims and questions, formulate hypotheses and make predictions
- identify, research and construct aims and questions for investigation
- identify independent, dependent and controlled variables in controlled experiments
- formulate hypotheses to focus investigations

Plan and conduct investigations
- determine appropriate investigation methodology: case study; classification and identification; controlled experiment (within subjects, between subjects, mixed design); correlational study; fieldwork; literature review; modelling; product, process or system development; simulation
- design and conduct investigations; select and use methods appropriate to the investigation, including consideration of sampling technique (random and stratified) and size to achieve representativeness, and consideration of equipment and procedures, taking into account potential sources of error and uncertainty; determine the type and amount of qualitative and/or quantitative data to be generated or collated

Comply with safety and ethical guidelines
- demonstrate ethical conduct and apply ethical guidelines when undertaking and reporting investigations

Construct evidence-based arguments and draw conclusions
- use reasoning to construct scientific arguments, and to draw and justify conclusions consistent with evidence base and relevant to the question under investigation
- discuss the implications of research findings and proposals, including appropriateness and application of data to different cultural groups and cultural biases in data and conclusions

Analyse, evaluate and communicate scientific ideas
- discuss relevant psychological information, ideas, concepts, theories and models and the connections between them

This activity will strengthen your research methodologies skills and reinforce your understanding of the effects of stress on the human body. Read the information below and answer the questions that follow.

Stress during pregnancy related to children's later movement, coordination

A longitudinal study by researchers at the University of Notre Dame Australia and the Telethon Kids Institute has found that stress experienced by mothers during pregnancy is related to their children's behaviour, as well as mental and cognitive outcomes in middle childhood and into adolescence. The study, which looked at the relationship between maternal pregnancy stress and children's motor development, found that mothers who experienced more stressful events during their pregnancies had children who scored lower on a test of movement competence.

To test the relationship between maternal stress and children's motor development, researchers followed 2900 primarily Caucasian Australian mothers. When the women were 18 weeks pregnant, they were asked to complete a questionnaire about stressful events during their pregnancies. These events included financial hardship, losing a close relative or friend, separation or divorce, marital problems, problems with the pregnancy, losing a job and moving residences. The mothers completed the same questionnaire when they were 34 weeks pregnant.

When the children born of those pregnancies were 10, 14 and 17 years old, they were assessed on their overall motor development (fine motor skills and gross motor skills) and coordination using a 10-item movement test. The test measured children's hand strength as well as their ability to touch a finger to one's nose and then back to the index finger, distance jump, walk along a line heel to toe and stand on one foot. The test also measured their ability to move small beads from one box to another, thread beads onto a rod, tap a finger over 10 seconds, turn a nut onto a bolt and slide a rod along a bar as slowly as possible. Children were grouped according to those born to mothers who experienced no stress during pregnancy, those born to mothers who experienced fewer than three stressful events during their pregnancies (low stress) and those born to mothers who experienced three or more stressful events during pregnancy (high stress).

The study found that children born to mothers who experienced more stressful events during pregnancy scored lower on motor development across all three survey years (ages 10, 14 and 17). This may suggest an accumulative effect of stress on the developing foetal motor system. The greatest differences in motor development outcomes were between individuals whose mothers experienced no stress and those who experienced high stress. Stressful events experienced in later pregnancy had more influence on children's motor development scores than those experienced earlier. According to the researchers, this may be related to the development of the cerebellar cortex, a part of the brain that develops later in pregnancy and that controls many motor outcomes.

Low motor development has been linked to poorer short- and long-term mental and physical health outcomes, so it is important to assess the early risk factors to provide early intervention and support. Children with low motor competence can have difficulty in everyday life with fine and gross motor tasks such as writing, throwing and running. However, with intervention and support, this can be improved in a number of cases.

'Given our findings on the importance of mothers' emotional and mental health on a wide range of developmental and health outcomes, programs aimed at detecting and reducing maternal stress during pregnancy may alert parents and health professionals to potential difficulties and improve the long-term outcomes for these children,' notes Beth Hands, professor of human movement at the University of Notre Dame Australia, who co-authored the study.

Source: Adapted from Society for Research in Child Development. (2015, October 14). Stress during pregnancy related to children's later movement, coordination. available at: sciencedaily, retrieved December 31, 2015 from www.sciencedaily.com/releases/2015/10/151014084812.htm

1. What was the aim of this study?

2. What is a longitudinal study? Identify one advantage and one limitation of this type of study.

3. What data collection methods were used in this study?

4. Identify the independent variable(s).

5. Identify the dependent variable(s).

6. Construct a possible hypothesis for this study.

7. Identify the control group and the experimental group(s) in this study.

8. What were the results of this study?

9. What conclusions can be drawn from these results?

10. Explain one major limitation of this investigation.

Exam practice

Multiple choice

Circle the response that best answers the question.

1. A _____ involves the physiological and psychological changes that people experience when they are confronted by a stressor.
   - A eustress reaction
   - B stress response
   - C primary appraisal
   - D secondary appraisal

Use the following information to answer Questions 2–4.

James was very happy when he and his wife bought their new house. A few weeks after the purchase, James was unable to work after contracting long COVID. At the time, his wife didn't have a paying job, so he didn't know how he would meet his mortgage repayments.

2. When he purchased his house, James immediately experienced
   - A distress.
   - B eustress.
   - C the freeze response.
   - D the flight-or-fight-or-freeze response.

3. Using the terminology of Lazarus and Folkman's Transactional Model of Stress and Coping, James' immediate appraisal of buying a new house was probably
   - A a threat.
   - B stressful.
   - C irrelevant.
   - D benign–positive.

4. Using the terminology of Lazarus and Folkman's Transactional Model of Stress and Coping, James' primary appraisal of his inability to work was that it was
   - A a threat.
   - B harm–loss.
   - C a challenge.
   - D benign–positive.

5. Chronic stress refers to the body's response to a(n) _____ stressor. Acute stress refers to the body's response to a(n) _____ stressor.
   - A internal; external
   - B persistent; immediate
   - C external; internal
   - D immediate; persistent

6. A stressful experience is:
   - A subjective in nature.
   - B objective in nature.
   - C a positive psychological response.
   - D caused only by external events.

7. Which one of the following statements about the stress response is correct?
   - A It is a controlled response when the sympathetic nervous system is activated.
   - B It is an automatic response when the sympathetic nervous system is activated.
   - C It is a controlled response when the parasympathetic nervous system is activated.
   - D It is an automatic response when the parasympathetic nervous system is activated.

8 The activation of the body's physiological systems to deal with a potential threat is called
   A the exhaustion stage.
   B the resistance stage.
   C stress.
   D the flight-or-fight-or-freeze response.

9 Dysbiosis is
   A disbelief in the presence of the microbiome.
   B an imbalance in the microbiome, causing disease.
   C when the microbiome is balanced and the individual is healthy.
   D when the microbiome stops working.

10 Which of the following are physiological responses of the body when the sympathetic nervous system is activated?
   A dilation of the pupils, increased heart rate and blood pressure
   B contraction of the pupils, decreased heart rate and contraction of the bronchioles
   C dilation of the bronchioles, blood channelled to the internal organs and increased digestion
   D contraction of the bronchioles, blood channelled to the internal organs and increased digestion

11 According to the GAS, countershock is the result of
   A the activation of the parasympathetic nervous system.
   B the activation of the sympathetic nervous system.
   C the release of cortisol by the liver and other internal organs.
   D the decrease in blood pressure and body temperature.

12 According to the GAS, during which of the following stages do signs of illness appear?
   A alarm–reaction stage
   B shock/countershock stage
   C stage of resistance
   D stage of exhaustion

13 Marion is walking down the street when she hears a loud bang and feels the ground shift beneath her feet. She realises that a bomb has gone off close by. She feels sick and can't decide which way to run. The best description of what Marion has experienced is
   A primary appraisal.
   B secondary appraisal.
   C the flight-or-fight-or-freeze response.
   D distress.

14 Prolonged activation of physiological systems as the result of a stressor
   A occurs only because of the presence of external stimuli.
   B occurs only in the alarm–reaction stage of the GAS.
   C can improve our performance on tasks involving cognitive manipulation of information.
   D can deplete the body's resources and lead to long-term illness or disease and/or psychological difficulties.

15 On his way to meet a friend, Solomon witnessed a car accident. He found that, because the accident involved people he did not know and no one was injured, it was not a very stressful situation for him. According to Lazarus and Folkman's Transactional Model of Stress and Coping, Solomon's reaction is an example of
   A a shock/countershock reaction.
   B coping flexibility.
   C a primary appraisal.
   D a secondary appraisal.

16 Which of the following statements about cortisol is correct?
   A Cortisol allows the body to deal with short-term stressors.
   B Secretion of cortisol is highest in the exhaustion stage of the GAS.
   C Cortisol is a hormone, secreted from the adrenal glands in response to stress.
   D The flight-or-fight-or-freeze response is driven by cortisol.

17 Which of the following is an example of primary appraisal according to Lazarus and Folkman's Transactional Model of Stress and Coping?
  A determining the extent to which additional resources are needed to cope
  B evaluating the potential impact of the stressor
  C judging the usefulness of coping resources that are available
  D any exchange between the individual and their environment

18 Which of the following is an example of secondary appraisal according to Lazarus and Folkman's Transactional Model of Stress and Coping?
  A making a judgement about whether a situation is actually stressful
  B minimising harm or loss that may occur
  C estimating the value of coping options and resources that may be used
  D minimising harm or loss that has occurred

19 An example of problem-focused coping is
  A venting.
  B wishful thinking.
  C escape–avoidance.
  D redefining the stressor in a more manageable way.

20 Ali wanted to conduct an experiment on the effect that physical exercise has on the body. He asked 10 friends and family members to be his participants. They all agreed. What sampling method did Ali use?
  A convenience sampling
  B stratified sampling
  C random sampling
  D random stratified sampling

Short answer

1 Andy is experiencing stress in response to news he has just received about his job. The company he is working for is experiencing a short-term financial problem and four staff are going to be made redundant. Andy has just got married and bought a new house.

  a Identify the stressor in this situation and state whether it is internal or external. 2 marks

  _____
  _____

  b State two physiological and two psychological responses that Andy could have in response to this stressor. 4 marks

  _____
  _____
  _____
  _____

2 Figure 3.9 shows the fluctuations in cortisol levels over the day.

Figure 3.9

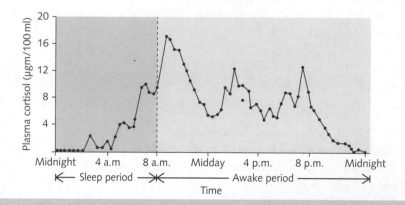

a Where in the body is cortisol produced? 1 mark

b State three times during the day that this person could be displaying stress symptoms. 3 marks

c What two physiological changes in the body would you predict will occur as a result of the high cortisol levels at these three times of the day? 2 marks

3 What is the gut–brain axis? 1 mark

4 How does the gut–brain axis affect your mental health? 4 marks

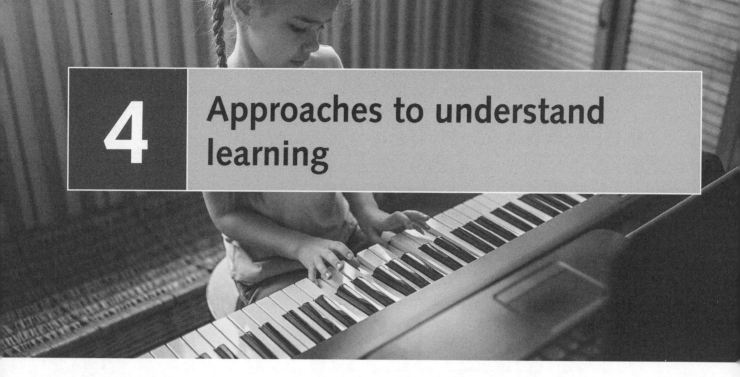

4 Approaches to understand learning

4.1 Approaches to understand learning

Key knowledge
- behaviourist approaches to learning, as illustrated by classical conditioning as a three-phase process (before conditioning, during conditioning and after conditioning) that results in the involuntary association between a neutral stimulus and unconditioned stimulus to produce a conditioned response, and operant conditioning as a three-phase process (antecedent, behaviour and consequence) involving reinforcement (positive and negative) and punishment (positive and negative)

4.1.1 Different approaches

Key science skills
Analyse, evaluate and communicate scientific ideas
- discuss relevant psychological information, ideas, concepts, theories and models and the connections between them

1 Define 'learning'.
___
___
___

2 This chapter explores three different approaches to learning. Each approach looks at learning in a different way. Define each approach below.

a Behaviourist
___
___
___

b Social cognitive
___
___
___

This activity will help you understand the structure of this chapter and the three different approaches to learning that are investigated.

c   Situated systems

_____
_____
_____

4.2 Behaviourist approaches to learning

Key knowledge
- behaviourist approaches to learning, as illustrated by classical conditioning as a three-phase process (before conditioning, during conditioning and after conditioning) that results in the involuntary association between a neutral stimulus and unconditioned stimulus to produce a conditioned response, and operant conditioning as a three-phase process (antecedent, behaviour and consequence) involving reinforcement (positive and negative) and punishment (positive and negative)

4.2.1 Classical conditioning process

Key science skills
Analyse, evaluate and communicate scientific ideas
- use appropriate psychological terminology, representations and conventions, including standard abbreviations, graphing conventions and units of measurement
- discuss relevant psychological information, ideas, concepts, theories and models and the connections between them
- analyse and explain how models and theories are used to organise and understand observed phenomena and concepts related to psychology, identifying limitations of selected models/theories

Develop

This activity will aid your understanding of Pavlov's experiments on classical conditioning by helping you to become familiar with the classical conditioning process and the definitions of classical conditioning terms. You will then apply the theory of classical conditioning to different situations.

PART A

1   Figure 4.1 illustrates the classical conditioning process. Complete the figure by filling in the blanks along with the explanations of the elements of classical conditioning.

Remember that before the association has been made, the stimulus and response are unconditioned. Once the association has been made, the stimulus and response are conditioned.

Figure 4.1 The classical conditioning process

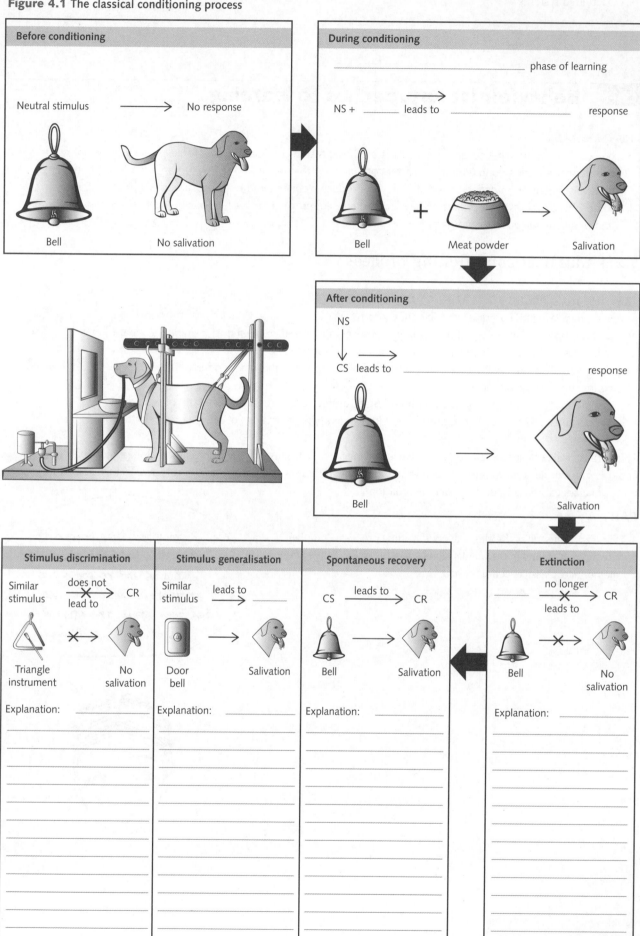

PART B

Table 4.1 contains a number of terms associated with classical conditioning. Choose terms from this table to answer the 'What am I?' and 'Who am I?' questions that follow.

Table 4.1 Terms associated with classical conditioning

Extinction	Neutral stimulus	Classical conditioning	Unconditioned response
Ivan Pavlov	Spontaneous recovery	Conditioned response	Unconditioned stimulus
Conditioned stimulus	Acquisition		

Each term is used only once.

1 I am a form of learning in which two unrelated stimuli are paired repeatedly so that existing reflex responses are elicited by a new stimulus. What am I?

2 I am a stimulus that evokes a response due to learning. What am I?

3 I am a stimulus that does not naturally elicit a response. What am I?

4 I am the automatic response to a specific unconditioned stimulus. What am I?

5 I occur when the conditioned stimulus no longer elicits a response due to the unconditioned stimulus being removed. What am I?

6 I am a specific stimulus that is innately capable of eliciting a reflex response. What am I?

7 I am the response that has been learnt. What am I?

8 I am the phase or period of time during which a response or behaviour is learnt. What am I?

9 I am the reappearance of the conditioned response after a period of extinction. What am I?

10 I am the famous researcher who was conducting experiments on the digestive systems of dogs when I discovered classical conditioning. Who am I?

4.2.2 Classical conditioning key terms

Key science skills
Analyse, evaluate and communicate scientific ideas
- use appropriate psychological terminology, representations and conventions, including standard abbreviations, graphing conventions and units of measurement
- discuss relevant psychological information, ideas, concepts, theories and models and the connections between them
- analyse and explain how models and theories are used to organise and understand observed phenomena and concepts related to psychology, identifying limitations of selected models/ theories

Develop

Some very specific key terms are used to describe the process of classical conditioning. You need to know these key terms and be able to apply them to new situations. This activity will help you to do this.

1. Read each scenario carefully.
2. For each scenario, identify the:
   - neutral stimulus (NS)
   - unconditioned stimulus (UCS)
   - unconditioned response (UCR)
   - conditioned stimulus (CS)
   - conditioned response (CR).
3. For each scenario, organise the events described into the three-phase process of classical conditioning using the table provided.

Take the time to get to know these terms. It will make things easier for you to understand.

SCENARIO 1

A few weeks ago, David took his girlfriend to a local restaurant. Over dinner, they had a serious disagreement, which resulted in David's girlfriend breaking up with him. They are now back together, but whenever David returns to the same restaurant with his girlfriend, he feels very nervous and gets butterflies in his stomach and sweaty palms.

NS: _____
UCS: _____
UCR: _____
CS: _____
CR: _____

Phase 1: Before conditioning	
Phase 2: During conditioning	Phase 3: After conditioning

SCENARIO 2

Gary was standing at the supermarket checkout one day, listening to his favourite song on his smartphone. A car lost control in the carpark and crashed through the front window of the store, narrowly missing Gary and other people nearby. Gary was very upset, and for several days afterwards his hands shook and he found himself bursting into tears without provocation. A week later, after the symptoms had gone, Gary was lying on the couch at home listening to the radio when his favourite song was played. His hands began to shake and he started crying.

NS: _____
UCS: _____
UCR: _____
CS: _____
CR: _____

Phase 1: Before conditioning	
Phase 2: During conditioning	Phase 3: After conditioning

SCENARIO 3

Sri takes her dog Strider for a walk every morning. During the summer months, Sri puts on a red baseball cap before going for the walk. Recently, Sri has noticed that each time she puts on the red cap, Strider wags his tail and stands at the front door.

NS: _____
UCS: _____
UCR: _____
CS: _____
CR: _____

Phase 1: Before conditioning	
Phase 2: During conditioning	Phase 3: After conditioning

SCENARIO 4

Flora has a cat named Tiger. Flora's routine when she comes home from work is to put her keys on the kitchen counter (which makes a clanging noise each time), then prepare Tiger's dinner. After several days of this routine, Flora notices that Tiger runs up to her and salivates whenever she puts her keys on the counter.

NS: _____
UCS: _____
UCR: _____
CS: _____
CR: _____

Phase 1: Before conditioning	
Phase 2: During conditioning	Phase 3: After conditioning

SCENARIO 5

When Paul was five years old, he was sitting on the steps at a house where his parents were visiting friends. There were three other children in the yard at the time, playing with the family dog. The children playing with the dog thought it might be fun to tie ribbons in its fur and dress it in a shirt they found on the clothesline. The dog did not want to be dressed up, so it ran to the house, where Paul was sitting, minding his own business while waiting for his parents. The other children followed the dog, and continued trying to dress it in the shirt. The dog started to bark, then turned and bit Paul on the hand. Paul screamed loudly. Paul's parents ran outside at the noise and found Paul holding his bleeding hand.

From then on, Paul was terrified of dogs. He felt anxious if he heard or saw a dog barking, and could not walk down a street where there was a dog in the front yard of a house. His grandparents had to lock their dog outside when Paul visited them. Paul would sit at the window and watch his sisters playing with the dog outside.

When Paul was 18 years old, he decided that he needed to overcome his extreme fear of dogs, because it was affecting his life. He wanted to travel overseas, but thought he couldn't until he was able to control the fear, because he might find himself in unexpected situations where dogs may be present. Paul went to see a psychologist for assistance.

NS: _____
UCS: _____
UCR: _____
CS: _____
CR: _____

Phase 1: Before conditioning	
Phase 2: During conditioning	Phase 3: After conditioning

4.2.3 Ethics in learning research: 'Little Albert'

Key science skills
Comply with safety and ethical guidelines
- demonstrate ethical conduct and apply ethical guidelines when undertaking and reporting investigations

Analyse, evaluate and communicate scientific ideas
- use appropriate psychological terminology, representations and conventions, including standard abbreviations, graphing conventions and units of measurement
- discuss relevant psychological information, ideas, concepts, theories and models and the connections between them

Watson and Rayner's 'Little Albert' experiment is famous for all the wrong reasons. Modern ethical guidelines that now govern psychological research mean that this experiment could not be conducted today.

PART A

1. Read the information in your textbook about Watson and Rayner's conditioning experiment involving Little Albert.
2. Fill in the blanks in Table 4.2 by identifying the conditioning elements involved in the experiment.

Table 4.2 Elements involved in Watson and Rayner's Little Albert experiment

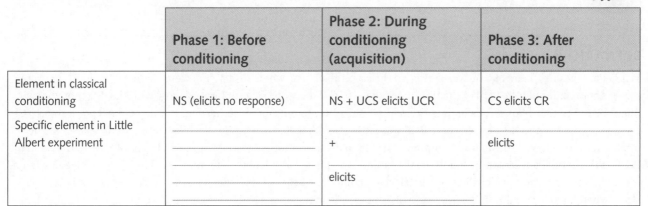

	Phase 1: Before conditioning	Phase 2: During conditioning (acquisition)	Phase 3: After conditioning
Element in classical conditioning	NS (elicits no response)	NS + UCS elicits UCR	CS elicits CR
Specific element in Little Albert experiment		_____ + _____ elicits _____	_____ elicits _____

PART B

1. Define each of the following ethical guidelines and give an example of how each was breached during Watson and Rayner's experiment.

   a. Informed consent procedures

   i. Definition

   ii. Example

b Debriefing
   i Definition

   ii Example

c Confidentiality
   i Definition

   ii Example

d Withdrawal rights
   i Definition

   ii Example

e Voluntary participation
   i Definition

   ii Example

2. Watson and Rayner's experiment has been criticised for being unethical for many reasons, including that they failed to properly extinguish the conditioned response. Outline a method Watson and Rayner could have used to extinguish the conditioned fear response in Little Albert.

4.2.4 Operant conditioning

Key science skills
Analyse, evaluate and communicate scientific ideas
- use appropriate psychological terminology, representations and conventions, including standard abbreviations, graphing conventions and units of measurement
- discuss relevant psychological information, ideas, concepts, theories and models and the connections between them

This activity has been designed to increase your understanding of the elements of operant conditioning.

> Operant conditioning is all your actions being followed by a reinforcement. Remember that the reinforcement can be either positive or negative.

PART A

1 Define 'positive reinforcement' and provide an example.

_____
_____
_____

2 Define 'negative reinforcement' and provide an example.

_____
_____

3 Define 'positive punishment' and provide an example.

_____
_____

4 Define 'negative punishment' and provide an example.

_____
_____

5 Describe the similarities and differences between positive reinforcement and negative reinforcement.

_____
_____
_____
_____
_____

6 Describe the similarities and differences between positive punishment and negative punishment.

_____
_____
_____
_____
_____

PART B

Table 4.3 contains descriptions of situations that are associated with the elements of operant conditioning. Tick the appropriate column to indicate which element of operant conditioning is being used in each situation: positive reinforcement, negative reinforcement, positive punishment or negative punishment.

Table 4.3 Examples of operant conditioning

Situation	Positive reinforcement	Negative reinforcement	Positive punishment	Negative punishment
A rat quickly learns to press a bar to stop an electric shock being administered through the floor of its cage.				
Nadia receives pocket money for helping around the house.				
Hannah crashes her parents' car into the garage after being told not to drive the car, so she is grounded for a month.				

Situation	Positive reinforcement	Negative reinforcement	Positive punishment	Negative punishment
Harry is fined $200 for speeding.				
Sam is talking in class and not doing any work, so the teacher walks to his desk and stands behind him. Sam stops talking and starts to work.				
Bridie receives an A+ on her Psychology SAC after studying hard.				
Violet's parents chastise her when she eats with her fingers at the dining table.				
Fido's owner puts a collar on him that releases an unpleasant sound that only dogs can hear every time he barks.				
I have a headache, so I take an aspirin and the headache goes away.				
Brian does his homework to stop his parents from confiscating his mobile phone.				
The judge tells Mr Axe that he is to spend 30 years in jail for the murder he committed.				
Ari earns a bonus at work for selling more computers than every other sales person.				
Sonia rubs some after-sun moisturiser on her sunburn and it stops itching. The next time she is sunburned, she applies the same moisturiser.				
Eric fails to return home from a party until after his curfew, so his parents ground him for a week.				
Mrs Garcia gives all of her students free time on the computers when they complete their work.				
Marvin is sent a bill for his mobile phone, but he forgets to pay it, so the company sends him another bill, which includes a $40 late fee.				

4.2.5 The ABCs of operant conditioning

Key science skills
Analyse, evaluate and communicate scientific ideas
- use appropriate psychological terminology, representations and conventions, including standard abbreviations, graphing conventions and units of measurement
- discuss relevant psychological information, ideas, concepts, theories and models and the connections between them

Develop

For each of the following scenarios, use the table provided to show the three-phase process of operant conditioning.

SCENARIO 1

Bernie has a job in which he is offered physical incentives, such as gifts, vouchers for restaurants and money bonuses, in return for reaching his sales targets on time. This motivates Bernie to perform better so that he can continuously get his incentives and bonus.

Antecedents	Behaviour	Consequences

The tables are provided to help you organise your thoughts.

SCENARIO 2

Jilly throws a tantrum because she didn't get the chocolate bar she wanted at the supermarket. Her father eventually gives in and gets her one. She then stops her tantrum. Next time she goes to the supermarket with her father she throws another tantrum and is again given a chocolate bar.

Antecedents	Behaviour	Consequences

SCENARIO 3

After hitting his classmate in English class, Cory is made to sit alone at the front of the class. No one is allowed to talk to him or sit with him.

Antecedents	Behaviour	Consequences

4.2.6 Using operant conditioning in animal training

Key science skills
Analyse, evaluate and communicate scientific ideas
- use appropriate psychological terminology, representations and conventions, including standard abbreviations, graphing conventions and units of measurement
- discuss relevant psychological information, ideas, concepts, theories and models and the connections between them
- analyse and explain how models and theories are used to organise and understand observed phenomena and concepts related to psychology, identifying limitations of selected models/theories

Develop

For each of the following scenarios about using operant conditioning in animal training, use the table provided to show the three-phase process of operant conditioning.

SCENARIO 1

Christina is an animal trainer at a marine park. She wants to train one of her dolphins to come to the edge of the pool when she blows a whistle. Normally, the dolphin comes to the edge when Christina kneels beside the pool and reaches her hand under the water and pats the dolphin.

Antecedents	Behaviour	Consequences

SCENARIO 2

Meryl wants to train her dog to sit before it crosses a road. She wants the sign to sit to be when she claps her hands.

Antecedents	Behaviour	Consequences

4.2.7 The psychology of advertising

Key science skills

Analyse, evaluate and communicate scientific ideas
- critically evaluate and interpret a range of scientific and media texts (including journal articles, mass media communications, opinions, policy documents and reports in the public domain), processes, claims and conclusions related to psychology by considering the quality of available evidence

Advertisers use the principles of classical and operant conditioning to sell product. Figure 4.2 shows an advertisement for a soft drink.

Figure 4.2 Advertisement designed to get you to buy a soft drink

Drink in the sunshine

1 Explain how the advertiser is using conditioning to get your attention and to get you to buy this product.

4.3 The social-cognitive approach to learning

Key knowledge
- social-cognitive approaches to learning, as illustrated by observational learning as a process involving attention, retention, reproduction, motivation and reinforcement

4.3.1 Observational learning

Key science skills
Analyse, evaluate and communicate scientific ideas
- use appropriate psychological terminology, representations and conventions, including standard abbreviations, graphing conventions and units of measurement
- discuss relevant psychological information, ideas, concepts, theories and models and the connections between them
- analyse and explain how models and theories are used to organise and understand observed phenomena and concepts related to psychology, identifying limitations of selected models/theories

Observational learning is learning by socialising, interacting with and observing others. This activity will strengthen your understanding of the stages of observational learning and help you apply them to a real-life situation.

PART A

Drew is a skilled football player who often kicks goals. To try to improve his playing, he watched a replay of last year's grand final game. At his next practice, he tried some of the techniques he had seen in the replay. In his next match, however, he scored fewer goals than he had in any game all season.

1 Identify the four stages of observational learning and apply them to this scenario.

Can you think of something you have learnt by watching others?

PART B

1 In your workbook (see Activity 4.3.2), textbook or on the Internet, read about Albert Bandura's 1960s Bobo doll experiments on observational learning.
2 Complete Table 4.4 by explaining each stage of the observational learning process in relation to Bandura's experiments.

Table 4.4 Bandura's experiments on observational learning

Stage in the observational learning process	Explanation in relation to Bandura's experiments
Attention	
Retention	
Reproduction	
Motivation and reinforcement	

4.3.2 Application of research methods

Key science skills
Develop aims and questions, formulate hypotheses and make predictions
- identify, research and construct aims and questions for investigation
- identify independent, dependent and controlled variables in controlled experiments

Plan and conduct investigations
- determine appropriate investigation methodology: case study; classification and identification; controlled experiment (within subjects, between subjects, mixed design); correlational study; fieldwork; literature review; modelling; product, process or system development; simulation

Comply with safety and ethical guidelines
- demonstrate ethical conduct and apply ethical guidelines when undertaking and reporting investigations

Construct evidence-based arguments and draw conclusions
- use reasoning to construct scientific arguments, and to draw and justify conclusions consistent with evidence base and relevant to the question under investigation

Use the following description of Bandura's experiments with Bobo dolls to improve your understanding of observational learning, particularly in children, and your understanding of research methods.

PARTICIPANTS

The participants involved in the study were 48 boys and 48 girls enrolled in the Stanford University Nursery School. They were aged from 2 to 5 years (35 to 69 months) and the mean age was 52 months.

An adult male and adult female served as models in both the real-life aggression and the filmed aggression conditions. A female experimenter conducted the study for all 96 children.

METHOD

There were three experimental groups with 24 children in each and one control group of 24 children. The experimental conditions consisted of participants observing:

- real-life aggressive behaviour by male and female adult models
- the same models and behaviour on film
- similar behaviour by an animated cartoon character with a high-pitched animated voice.

The experimental groups were further divided into male and female participants so that half the participants in condition one and two (observing the adult models) were exposed to same-gender models, while the other half were exposed to models of the opposite gender.

Following observation of the aggressive behaviour either by a human or cartoon character, the children were tested for the amount of imitative and non-imitative aggression in a different experimental setting without the presence of the models. Imitative aggression was those behaviours that replicated the modelled aggressive behaviour, and non-imitative aggression was those behaviours that the children exhibited without the behaviour having been modelled in the experiment. The control condition was 'no exposure to the aggressive models'.

Participants in the real-life aggressive condition were individually brought to an experimental room with the model, and were invited by the experimenter to join in a game. The participant was then led to a corner of the room and seated at a table that contained various games and activities. After demonstrating some of the activities, the experimenter led the adult model to a table in another corner of the room containing activities as well as a mallet and a 1.5-metre-tall inflatable Bobo doll. The experimenter informed the model that this was their play area, the model was seated and then the experimenter left the room.

The adult model spent approximately one minute playing with some of the activities at the table, then turned to the Bobo doll and spent the remainder of the time behaving aggressively towards it in novel ways that a child was unlikely to perform of their own accord.

The model exhibited the following distinct aggressive acts that were to be scored as imitative responses.

- The model sat on the doll and punched it repeatedly in the nose.
- The model raised the doll and hit it on the head with the mallet.
- The model tossed the doll in the air and kicked it around the room (three times).
- Throwing the doll and kicking it was interspersed with verbal aggression, such as 'sock him in the nose' and 'hit him down'.

The participants in the adult model film condition were shown to the same room as those in condition one, with the same activities, but instead of the adult model behaving aggressively in the same room, they were shown a film depicting the same adult models used in condition one, exhibiting identical aggressive behaviour to the first condition.

In the third condition, participants were treated in the same way as the previous two situations. However, they were shown a cartoon instead of a real-life model or a film of a human model. The film was titled *Herman the Cat* and depicted a cat behaving aggressively towards the Bobo doll in a similar manner to the human models.

After the participants had been exposed to the aggressive condition, the experimenter took them individually to another room, which held a variety of highly attractive toys. The experimenter explained that the participant could play with the toys. Once the participant was sufficiently engaged in play with the toys, the experimenter said that they were her very best toys and that she had decided to reserve them for other children. The participant was then led to another room that contained toys and was told they could play with those.

This new room was the experimental room and contained a variety of toys including a 1-metre-tall Bobo doll and a mallet. The participants spent 20 minutes in this room and their behaviour was observed through a one-way mirror.

The results from this experiment are shown in Table 4.5.

RESULTS

Table 4.5 Imitation of aggressive models

Response category	Experimental groups					Control group
	Real life		Film		Cartoon film	
	F model	M model	F model	M model		
Total aggression						
Girls	65.8	57.3	87.0	79.5	80.9	36.4
Boys	76.8	131.8	114.5	85.0	117.2	72.2
Imitative aggression						
Girls	19.2	9.2	10.0	8.0	7.8	1.8
Boys	18.4	38.4	34.3	13.3	16.2	3.9
Mallet aggression						
Girls	17.2	18.7	49.2	19.5	36.8	13.1
Boys	15.5	28.8	20.5	16.3	12.5	13.5
Sits on Bobo doll*						
Girls	10.4	5.6	10.3	4.5	15.3	3.3
Boys	1.3	0.7	7.7	0.0	5.6	0.6
Non-imitative aggression						
Girls	27.6	24.9	24.0	34.3	27.5	17.8
Boys	35.5	48.6	46.8	31.8	71.8	40.4
Aggressive gun play						
Girls	1.8	4.5	3.8	17.6	8.8	3.7
Boys	7.3	15.9	12.8	23.7	16.6	14.3

* This response category was not included in the total aggression score.

Source: Bandura, A., Ross, D. & Ross, S. A. (1963). Imitation of Film-mediated Aggressive Models. *Journal of Abnormal and Social Psychology*, 66(1), pp. 3–11. American Psychological Association.

1. Identify the methodology of this experiment.

2. What was the aim of this experiment?

3. What were the independent and dependent variables being tested?
   Independent variable _____
   Dependent variable _____

4. Write a discussion section for this experiment. Include the following information: hypothesis supported or rejected, possible extraneous variables, conclusions regarding the theory and possible generalisations.

5 Explain at least two ethical concepts or guidelines the experimenters would have needed to consider to conduct this experiment.

_____

_____

_____

4.4 Aboriginal and Torres Strait Islander peoples' approaches to learning

Key knowledge
- approaches to learning that situate the learner within a system, as illustrated by Aboriginal and Torres Strait Islander ways of knowing where learning is viewed as being embedded in relationships where the learner is part of a multimodal system of knowledge patterned on Country

4.4.1 The situated multimodal systems approach to learning

Key science skills
Analyse, evaluate and communicate scientific ideas
- discuss relevant psychological information, ideas, concepts, theories and models and the connections between them

PART A

Although there are many distinct languages and cultural practices among the many different Indigenous Australian cultural groups, there are some common elements that we will describe as the Indigenous Australian situated multimodal systems approach to learning.

1 In order to understand the situated multimodal systems approach to learning, it is best to start off by understanding what the name incorporates. Explain the name by stating what each adjective means in terms of the wider approach to learning. Use Figure 4.3 to assist you in organising your answer.

Figure 4.3 Defining terms

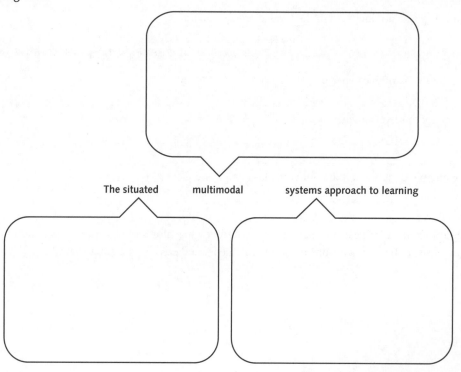

PART B

The situated multimodal systems approach to learning uses symbols to convey meaning both within and between generations. We sometimes use symbols to convey meaning when we are communicating with others.

1 State what each of the following Western symbols means to you.

a ;)

b ROTFL

c <3

d <(")

2 Figure 4.4 shows some symbols that are used by Indigenous Australian peoples to communicate meaning both within and between generations. See if you can work out what each of these symbols mean.

Figure 4.4 Examples of Indigenous Australian symbols

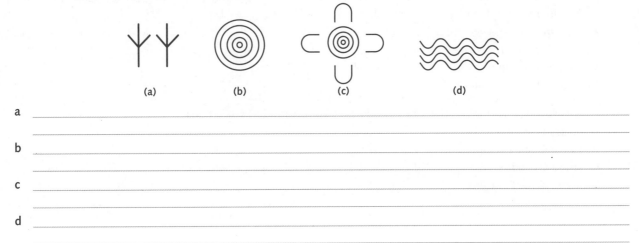

(a)   (b)   (c)   (d)

a

b

c

d

PART C

1 Match each term from the list below with its description in Table 4.6.

- Dreaming; Country; Ways of doing; Songlines; Ways of being; Ways of knowing

Table 4.6

Term	Description
	Travel routes that cross *Country* where Indigenous Australians use cues from the land to pass on knowledge, cultural values and wisdom.
	Indigenous Australian approaches to relational interactions between people and *Country*, including listening deeply to what *Country* has to teach.
	Indigenous Australian cultural practices that physically embody ways of knowing and being through performance, creating objects and engaging in caring for *Country*, including food, medicines and land management practices.
	Indigenous Australian approaches to learning and knowledge based on core understanding of the interrelationship between all entities, so that the learner and knowledge are embedded within a system of interrelationships.
	An Indigenous Australian understanding of place as a system of interrelated living entities, including the learner, their family, communities and interrelationships with land, sky, waterways, geographical features, climate, animals and plants.
	The body of ancestral knowledge that has been passed down over thousands of generations.

Exam practice

Multiple choice

Circle the response that best answers the question.

1. The researcher who is associated with the study of classical conditioning is
   A Albert Bandura.
   B B. F. Skinner.
   C John B. Watson.
   D Ivan Pavlov.

2. In a classical conditioning experiment, the dog salivating to a previously neutral stimulus is an example of a(n)
   A conditioned response.
   B unconditioned stimulus.
   C conditioned stimulus.
   D unconditioned response.

3. Every time Harriet's dog heard the sound of the electric can opener it began to salivate because it associated the sound with receiving food. In this example, the can opener is a(n)
   A conditioned response.
   B unconditioned response.
   C conditioned stimulus.
   D unconditioned stimulus.

4. Which learning theory states that the organism is learning through associations?
   A operant conditioning
   B instrumental conditioning
   C classical conditioning
   D observational learning

5. After Pavlov had conditioned his dogs to salivate to a tone, he repeatedly presented the tone without presenting the food. As a result, _____ occurred.
   A stimulus generalisation
   B stimulus discrimination
   C spontaneous recovery
   D extinction

6. If you were to blow a puff of air into a person's face, they would reflexively blink their eyes. According to classical conditioning, the puff of air would be termed the
   A unconditioned stimulus.
   B conditioned stimulus.
   C unconditioned response.
   D conditioned response.

7. Negative reinforcement _____ the likelihood of a behaviour being repeated and positive punishment _____ the likelihood of a behaviour being repeated.
   A increases; increases
   B increases; decreases
   C decreases; increases
   D decreases; decreases

8 Olive rubbed some cortisone cream on the spot where a mosquito had bitten her and the itching stopped. The next time a mosquito bit her, she reached for the cortisone cream. In terms of operant conditioning, this is an example of
   A positive reinforcement.
   B positive punishment.
   C negative reinforcement.
   D negative punishment.

9 Operant conditioning focuses on how
   A people learn through associations.
   B behaviour is influenced by observing those around us.
   C behaviour is influenced by the consequences that follow.
   D a change in behaviour is related to emotional events.

10 Compared to classical conditioning, the behaviours that are learned through operant conditioning are
   A reflexive.
   B involuntary.
   C elicited.
   D voluntary.

11 Observational learning occurs in what order?
   A motivation, reinforcement, attention, reproduction and retention
   B attention, reproduction, retention, motivation and reinforcement
   C attention, retention, reproduction, motivation and reinforcement
   D motivation, attention, retention, reproduction and reinforcement

12 Bridget's school calls her parents because she has been sending offensive text messages to her fellow students. The next day, Bridget's parents take her mobile phone away from her for a week. This is an example of
   A positive reinforcement.
   B negative reinforcement.
   C positive punishment.
   D negative punishment.

13 Which of the following must occur in both classical and operant conditioning for spontaneous recovery to take place?
   A The conditioned response must be extinguished.
   B There must be a rest period.
   C The conditioned response must reappear.
   D All of the above.

14 In classical conditioning, the learner is _____, while in operant conditioning the learner is _____.
   A voluntary; involuntary
   B involuntary; voluntary
   C passive; active
   D active; passive

15 In Bandura's original experiments on aggression and observational learning, he found that the number of aggressive acts displayed was at its highest when observing
   A a model act aggressively.
   B a model act pleasantly.
   C a model do nothing.
   D all of the above.

16 John Watson and Rosalie Rayner conducted a series of experiments on their subject 'Little Albert'. They tried to condition a(n) _____ response using _____ conditioning.
   A unconditioned; classical
   B fear; classical
   C reinforced; operant
   D negative; operant

17 Which ethical guideline was breached by Watson and Rayner in their work with 'Little Albert'?
   A non-maleficence
   B respect
   C justice
   D voluntary participation

18 Jeremy is 10 years old and learning to snow ski. He spends several hours watching skiers on television participating in the Winter Olympics and can't wait to be a champion himself. Due to Jeremy's lack of balance and coordination, he will most likely be unable to ski like an Olympian. Which important aspect of observational learning is missing from Jeremy's learning process?
   A reproduction
   B attention
   C retention
   D motivation

19 In Indigenous Australian culture, Country is considered to be
   A only made up of people and the land on which they live.
   B a living system of interrelationships between all living and non-living things.
   C a living system of interrelationships between all non-living things.
   D a living system of interrelationships between all living things.

20 Ways of doing enable Indigenous Australians to convey knowledge through
   A storytelling and dance.
   B writing and hunting.
   C recording and songs.
   D crafting objects and alphabet.

Short answer

1 Complete Table 4.7 to show the differences between classical conditioning and operant conditioning.   10 marks

Table 4.7

	Classical conditioning	Operant conditioning
What is it?		
Is the response voluntary or involuntary?		
How is it acquired?		
How is it extinguished?		
Is there spontaneous recovery?		

2 A mother is showing her young daughter how to make cupcakes. The mother is weighing all the ingredients and adding them to the bowl. The daughter is stirring the mixture. Next time they make cupcakes together the mother lets her daughter weigh the ingredients as well as add them to the bowl. List the four cognitive processes involved in this example of observational learning and explain how each one relates to the example given. 8 marks

3 Distinguish between positive punishment and negative punishment in operant conditioning. 4 marks

5 The psychobiological process of memory

5.1 What is memory?

Key knowledge
- the explanatory power of the Atkinson-Shiffrin multi-store model of memory in the encoding, storage and retrieval of stored information in sensory, short-term and long-term memory stores

5.1.1 Atkinson and Shiffrin's multi-store model of memory

Key science skills
Analyse, evaluate and communicate scientific ideas
- use appropriate psychological terminology, representations and conventions, including standard abbreviations, graphing conventions and units of measurement
- discuss relevant psychological information, ideas, concepts, theories and models and the connections between them

This activity will strengthen your understanding of the Atkinson-Shiffrin multi-store model of memory as you create a single-page summary of the model.
Use the terms in Table 5.1 to complete the paragraphs.

Some terms may be used more than once.

Table 5.1 Terms associated with memory

1/3–1/2 of a second	auditory	selective attention
18–20 seconds	attend	elaborative rehearsal
duration	long-term	maintenance rehearsal
echo	echoic sensory	registered
3–4 seconds	short-term	iconic sensory

Out of the corner of his eye, Jason noticed a sign in the window of his local supermarket, advertising for staff. This sensory information, which was being held in his _____ memory, was _____ in there for approximately _____.

Jason did not notice the 'Only females need apply' statement in the corner of the sign because he used _____, a process that filtered out this specific information.

Jason focused his attention on the contact telephone number for a few seconds as he wanted to apply for the job. This information was transferred to his _____ memory, where he could focus on it.

Jason knew this information would remain in his short-term memory for a limited time (approximately _____) and then drop out, unless he continued to _____ to it. He also knew that he must store the telephone number in his _____ memory if he wanted to retrieve it later. To avoid the number dropping out of his short-term memory, Jason began repeating it to himself while he searched his pockets for something on which to record the number. In other words, Jason chose to use _____ as a means of increasing the _____ of his memory.

As he was writing the telephone number on a piece of paper that he found, Jason realised that the first four digits were the same as his birth date and the second four were the same as the first four numbers of his bank account. By linking this new information with information already stored in his _____ memory, Jason now had a better chance of remembering the telephone number. He also remembered that he was using a technique called _____.

Jason's attention was drawn away from the sign when he heard his friend Jack call his name. This _____ information registered in his _____ memory and stayed there for approximately _____ in the form of a(n) _____.

5.2 The structure of long-term memory

Key knowledge
- the roles of the hippocampus, amygdala, neocortex, basal ganglia and cerebellum in long-term implicit and explicit memories

5.2.1 Concept map of memory

Key science skills
Analyse, evaluate and communicate scientific ideas
- discuss relevant psychological information, ideas, concepts, theories and models and the connections between them
- analyse and explain how models and theories are used to organise and understand observed phenomena and concepts related to psychology, identifying limitations of selected models/theories

Develop

Use the terms in Table 5.2 to complete the concept map in Figure 5.1.

Table 5.2 Terms associated with the multi-store model of memory

short-term	limited	chunking	declarative	rehearsed
unlimited	attended to	auditory	sensory	1/3–1/2 of a second
displaced	semantic	procedural	episodic	encoding
elaborative	consolidation	serial position effect	sensory	organisation
skills	3–4 seconds	attended to and rehearsed	long-term	18–20 seconds
iconic	visual			

Figure 5.1 Concept map of the Atkinson-Shiffrin multi-store model of memory

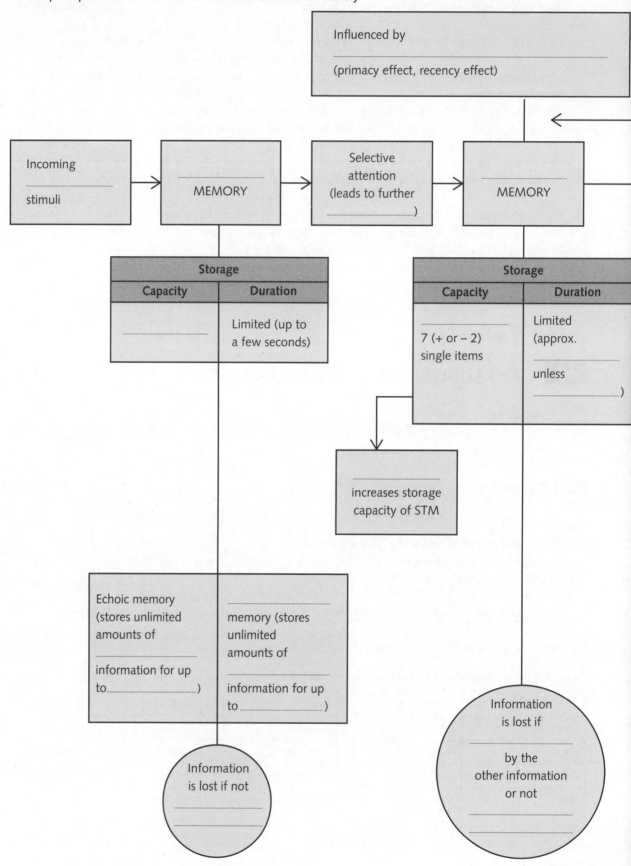

CHAPTER 5 / The psychobiological process of memory

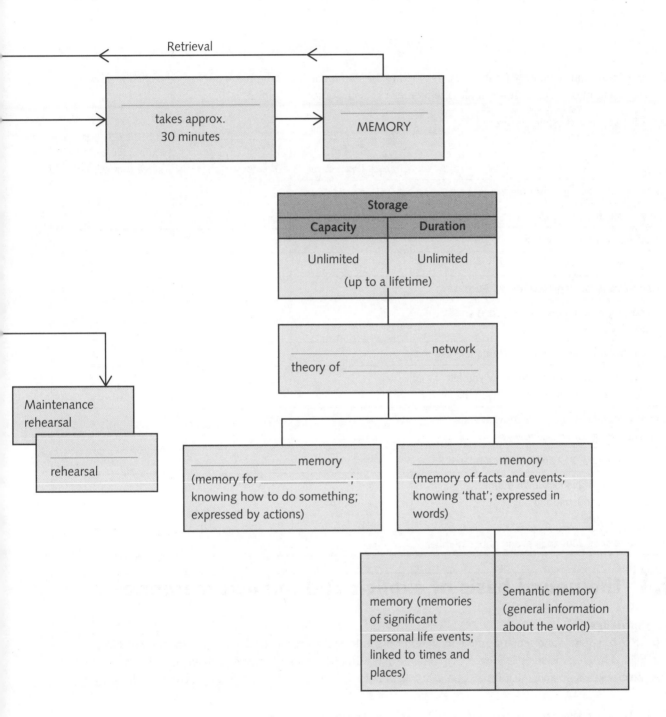

5.2.2 Comparing implicit and explicit memories

Key science skills
Analyse, evaluate and communicate scientific ideas
- discuss relevant psychological information, ideas, concepts, theories and models and the connections between them

Place a tick in the appropriate column to indicate whether the statement is an example of implicit or explicit memories. For explicit memories, decide if the example is semantic or episodic-autobiographical.

Table 5.3 Implicit and explicit memory

	Implicit (Procedural)	Explicit (Semantic)	Explicit (Episodic-autobiographical)
Knowledge of words and their meanings			
Tying your shoelaces			
Memories of your first day at school			
Sending a text message			
Knowing who is the Prime Minister of Australia			
Remembering what games you played at your 14th birthday party			
Being able to describe what an elephant looks like			
Memories of where you were and how you felt when you heard that your uncle had won the lottery			
Being able to translate into English what your Spanish-speaking cousin said to you			
Being able to add 2, 2, 2 and 2 to get a total of 8			
Remembering being in lockdown during the COVID-19 pandemic and doing schoolwork remotely			
Being able to state your address and phone number			

5.3 The neural basis of explicit and implicit memories

Key knowledge
- the role of episodic and semantic memory in retrieving autobiographical events and in constructing possible imagined futures, including evidence from brain imaging and post-mortem studies of brain lesions in people with Alzheimer's disease and aphantasia as an example of individual differences in the experience of mental imagery

5.3.1 Brain structures involved in long-term memory

Key science skills
Analyse, evaluate and communicate scientific ideas
- discuss relevant psychological information, ideas, concepts, theories and models and the connections between them

This activity will test your ability to differentiate between the different types of long-term memories and the major structures involved in long-term memory formation and storage.

PART A

Complete Figure 5.2.
1. Write a definition for long-term memory and its categories.
2. Describe the role of each of the structures involved in long-term memory.

Figure 5.2 Brain structures and concepts involved in long-term memory

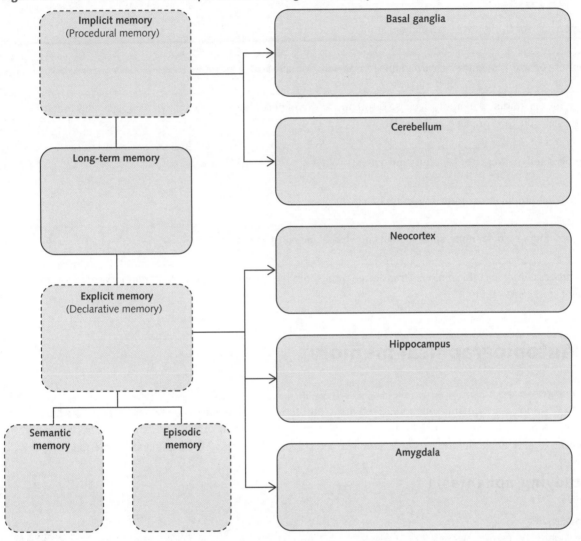

PART B

Use the terms in Table 5.4 to answer the following 'What am I?' questions.

Table 5.4 Terms and structures related to long-term memory formation

Basal ganglia	Neocortex	Hippocampus	Explicit memory
Cerebellum	Long-term memory	Episodic memory	Consolidation
Implicit memory	Amygdala	Semantic memory	

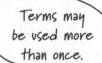
Terms may be used more than once.

1. If I am damaged, the result can be a loss of memory of facts and personally significant events. What am I?

2. I am the brain structure responsible for evaluating stimuli for danger or threat and determining the emotional significance of the stimuli. What am I?

3. I integrate and evaluate information sent from all lobes and I relay information to, and receive information from, other brain structures involved in memory formation and storage. What am I?

4. I am located deep within the brain and I extend into both temporal lobes. I am responsible for the formation and consolidation of explicit memories. What am I?

5. I am a structure attached to the rear of the brainstem responsible for the formation and storage of implicit memory of balance and coordination. What am I?

6. I am responsible for the relatively permanent storage of memories. What am I?

7. I am your unconscious memory of learned fine motor skills. What am I?

8. I am your autobiographical memory because I store memories of personally significant events or episodes that are related to a specific time or place. What am I?

9. I store procedural memories of voluntary movements. What am I?

10. I have cells that produce dopamine. What am I?

11. I am your memory store for impersonal factual knowledge and rules. What am I?

12. I am the process by which information is transferred from short-term memory to long-term memory for relatively permanent storage. What am I?

5.4 Autobiographical memory

Key knowledge
- the role of episodic and semantic memory in retrieving autobiographical events and in constructing possible imagined futures, including evidence from brain imaging and post-mortem studies of brain lesions in people with Alzheimer's disease and aphantasia as an example of individual differences in the experience of mental imagery

5.4.1 Studying aphantasia

Key science skills
Analyse, evaluate and communicate scientific ideas
- discuss relevant psychological information, ideas, concepts, theories and models and the connections between them
- critically evaluate and interpret a range of scientific and media texts (including journal articles, mass media communications, opinions, policy documents and reports in the public domain), processes, claims and conclusions related to psychology by considering the quality of available evidence
- acknowledge sources of information and assistance, and use standard scientific referencing conventions

Develop

This activity will help you to understand the planning involved when studying a new topic.
Firstly, read the information about aphantasia in your textbook, then complete the following tasks.

1. Your research question is 'What is it like to live with aphantasia?'. Write an aim relating to this research question.

2 Conduct a literature review of articles or reports that describe what aphantasia is like. For example, do people with aphantasia dream? What do they experience when reading a book? How would this differ if a book was read to them? Could someone with aphantasia write a fairy tale? Summarise two of these articles and include a full APA-style reference for each one. The summary should cover how a person with aphantasia describes a part of their experience that is different to people who don't have aphantasia.

ARTICLE 1

ARTICLE 1 REFERENCE

ARTICLE 2

ARTICLE 2 REFERENCE

3 Find out how aphantasia is diagnosed. Describe two tests that might be used.

5.5 Mnemonics

Key knowledge
- the use of mnemonics (acronyms, acrostics and the method of loci) by written cultures to increase the encoding, storage and retrieval of information as compared with the use of mnemonics such as sung narrative used by oral cultures, including Aboriginal peoples' use of Songlines

5.5.1 Mnemonics: acronyms and acrostics

Key science skills
Analyse, evaluate and communicate scientific ideas
- discuss relevant psychological information, ideas, concepts, theories and models and the connections between them

This activity will help you to understand mnemonic devices and how they can be used.

PART A

1 Define 'acronym'.

2 Find out what the following acronyms stand for.
   a DRSABCD

   b ASAP

   c AWOL

   d RSVP

   e EFTPOS

   f UNICEF

   g DOB

   h SCUBA

   i ABC (television station)

   j CEO

PART B

1 Define 'acrostic'.

2 Write an acrostic for the sections relating to observational learning: attention, retention, reproduction, motivation and reinforcement.
   _____
   _____
   _____
   _____

3 Compare the usefulness of acronyms and acrostics for a VCE student.
   _____
   _____
   _____
   _____

5.5.2 Mnemonics: method of loci

Key science skills
Analyse, evaluate and communicate scientific ideas
- discuss relevant psychological information, ideas, concepts, theories and models and the connections between them

This activity will help you to understand mnemonic devices and how they can be used.

PART A
Use the method of loci to create a strategy to remember the following words in order.
- multi-store
- sensory
- iconic
- echoic
- short-term
- maintenance rehearsal
- elaborative rehearsal
- chunking
- long-term
- retrieval

1 Describe a location. Include 10 loci in your description. Write the loci (e.g. letterbox, footpath) in the spaces provided below.
2 Place a word from the list above, in order, beside each of the 10 loci you have named.
3 Spend 10 minutes rehearsing the list. You can look at the list initially, but by the end of the 10 minutes, you should be able to recall the list without looking at the page.

Location (e.g. my backyard): _____

1 _____
2 _____
3 _____
4 _____
5 _____
6 _____
7 _____
8 _____
9 _____
10 _____

PART B

Aboriginal Songlines are used to create pathways to follow across Australia. Several roads in Melbourne started out as Aboriginal Songlines.

1 Name four roads or highways in or near Melbourne that were once Songlines.

_____
_____
_____
_____

5.5.3 Research into memory

> **Key science skills**
> Develop aims and questions, formulate hypotheses and make predictions
> - identify, research and construct aims and questions for investigation
> - identify independent, dependent and controlled variables in controlled experiments
> - formulate hypotheses to focus investigations
>
> Construct evidence-based arguments and draw conclusions
> - use reasoning to construct scientific arguments, and to draw and justify conclusions consistent with evidence base and relevant to the question under investigation
> - discuss the implications of research findings and proposals, including appropriateness and application of data to different cultural groups and cultural biases in data and conclusions
>
> Analyse, evaluate and communicate scientific ideas
> - discuss relevant psychological information, ideas, concepts, theories and models and the connections between them

Develop

This activity will help you to understand the work of a famous researcher, Elizabeth Loftus, who carried out several studies on the validity of eyewitness testimony in a court of law.

PART A

Read the passage and then answer the questions that follow.

> In a study by Elizabeth Loftus in 1975, 150 students at the University of Washington viewed a brief video of a car accident and then answered 10 questions about the accident. The critical question was about a white sports car (Loftus, 1975).
>
> Half of the subjects were asked, 'How fast was the white sports car going while travelling along the country road?', and half were asked, 'How fast was the white sports car going when it passed the barn while travelling along the country road?'. There was, in fact, no barn in the scene at all. Once they had completed the questionnaire, participants were free to go, but they were asked to return for some follow-up questions a week later.
>
> All of the subjects returned one week later and, without reviewing the video, filled in a second questionnaire. The final question was, 'Did you see a barn?', and subjects responded by circling either 'Yes' or 'No'.
>
> Of the subjects who were initially exposed to the false presupposition of the barn (i.e. the group that was asked how fast the car was going when it passed the barn), 17.3 per cent responded 'Yes' to the final question. Of the subjects who were not exposed to the false presupposition of the barn (i.e. the group that was simply asked how fast the car was going, with no mention of the barn), only 2.7 per cent claimed on their questionnaires to have seen the barn.
>
> It was concluded that an initial question containing a false presupposition can influence a witness's later tendency to report the presence of the non-existent object corresponding to that presupposition.

1 What was the sample in this study?

_____

2 Would you consider this sample to be representative? Explain your answer.

_____
_____
_____

3 Would you classify this study as a controlled experiment? Explain your answer.

4 Which group was the control group and which group was the experimental group in this study?

5 State the independent variable in this study.

6 State the dependent variable in this study.

7 Write a possible hypothesis for the study.

8 Graph the findings of this study on the axes in Figure 5.3. Make sure you add a scale to each axis and write a title for the graph.

Figure 5.3 Results

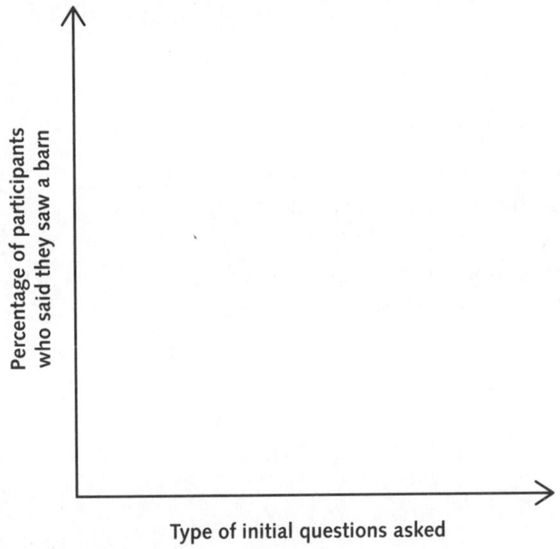

y-axis: Percentage of participants who said they saw a barn
x-axis: Type of initial questions asked

9 Do you believe that the difference between the two groups is significant? Explain your answer.

10 Do you believe that the conclusion for this study was valid? Explain your answer.

11 Identify one major limitation of this investigation.

Exam practice

Multiple choice

Circle the response that best answers the question.

1. Atkinson and Shiffrin's multi-store model of memory suggests that
   A there are three levels of memory: sensory, short-term and long-term.
   B there are different locations in the brain and central nervous system that process the different forms of information.
   C the brain acts like a computer; it decides what information is to be processed and what information to ignore.
   D we are programmed to act in certain ways; for example, sensory memory is primed to respond to movement and change.

2. When information is processed in memory, which three functions are performed?
   A sensory memory, short-term memory, long-term memory
   B recall, recognition, retrieval
   C encoding, storage, retrieval
   D maintenance rehearsal, elaborative rehearsal, chunking

3. Which of the following determines what information moves from sensory memory to short-term memory?
   A consolidation
   B primacy effect
   C selective attention
   D rehearsal

4. Which of the following statements about storage in long-term memory is true?
   A Information enters one of two storage areas, according to whether it is personal or objective knowledge.
   B Information in long-term memory is organised chronologically; that is, according to the date that it was learned.
   C Information in long-term memory is stored in terms of the physical qualities of the experience.
   D Information in long-term memory is believed to be organised in the form of semantic networks.

5. Information in sensory memory is available for
   A approximately 1/3–1/2 of a second.
   B up to a few seconds.
   C approximately 18–20 seconds, unless rehearsed.
   D a very long time, even up to a lifetime.

6. Which type of memory stores information about words like 'steam engine', including their meaning?
   A implicit memory
   B procedural memory
   C explicit episodic memory
   D explicit semantic memory

7. A typical multiple-choice question on a Psychology exam is an example of both a(n) _____ and a(n) _____ test of memory.
   A recall; implicit
   B recall; explicit
   C recognition; implicit
   D recognition; explicit

8 Eighty-two-year-old Mavis often has arguments with her daughter, Margaret, about who was present at her 80th birthday party. For instance, Mavis insists that her eldest son, Donald, was there. However, Margaret knows that Donald was overseas at the time. This indicates that Mavis's _____ memory has been impaired.
   A short-term
   B procedural
   C explicit semantic
   D explicit episodic

9 Free recall is
   A retrieving information from short-term memory with no cues.
   B retrieving information from long-term memory with no cues.
   C retrieving information from short-term memory using retrieval cues.
   D retrieving information from long-term memory using retrieval cues.

10 Amy, a 23-year-old university student, suffered a brain injury and sustained damage to her amygdala. Amy is most likely to experience difficulty with
   A implicit memory.
   B explicit memory.
   C semantic memory.
   D procedural memory.

11 Which of the following is an accurate description of elaborative rehearsal?
   A Elaborative rehearsal involves rote learning.
   B Elaborative rehearsal involves repetition of information vocally.
   C Elaborative rehearsal involves the mental repetition of information.
   D Elaborative rehearsal relies on making meaningful connections with information to be learnt.

12 Luke is learning to shoot a rifle and must remember the steps correctly. He came up with 'BRASS' to help him remember breathe, relax, aim, sight, squeeze. Luke is using:
   A a mnemonic device called an acronym.
   B a mnemonic device called an acrostic.
   C a retrieval cue.
   D implicit memories.

13 Which of the following statements is correct?
   A Iconic and echoic memory are parts of long-term memory.
   B Iconic and echoic memory are parts of short-term memory.
   C Iconic sensory memory relates to vision or touch; echoic sensory memory relates to hearing.
   D Iconic sensory memory relates to vision; echoic sensory memory relates to hearing.

14 Which statement about aphantasia is correct?
   A People with aphantasia will often daydream.
   B Aphantasia is also known as mind blindness.
   C People with aphantasia are unable to draw or paint pictures.
   D People with aphantasia can create mental images of people and places if they are familiar.

15 Which one of the following structures in the brain has a role in memory formation?
   A prefrontal cortex
   B neocortex
   C visual cortex
   D auditory cortex

16 Damage to the cerebellum from an accident is likely to affect
   A   the ability to use mnemonics.
   B   explicit memory formation.
   C   implicit memory retrieval.
   D   the ability to use chunking.

17 The two major causes of forgetting are
   A   retrieval failure and elaborative encoding.
   B   retrieval failure and interference.
   C   elaborative encoding and interference.
   D   the recency effect and the primacy effect.

18 The inability to store new memories for events and facts is known as
   A   anterograde amnesia.
   B   retrograde amnesia.
   C   implicit autobiographical memory loss.
   D   retrieval failure.

19 Which one of the following would not be considered to be episodic future thinking?
   A   Mentally repeating a phone number over and over until you use it.
   B   Imagining walking onto the stage to receive an award for excellence.
   C   Sitting in class and thinking of your walk home from school.
   D   Imagining being a parent for the first time.

20 Which brain areas are responsible for long-term potentiation (LTP)?
   A   hippocampus and amygdala
   B   amygdala and basal ganglia
   C   neocortex and basal ganglia
   D   hippocampus and neocortex

Short answer

1 Clearly distinguish between the various types of memory in the Atkinson-Shiffrin multi-store model.   3 marks

2 Describe the processes of encoding, storage and retrieval.   3 marks

3 a Name the two areas of the brain responsible for procedural memories. 2 marks

   b Explain the roles of each of these brain areas. 2 marks

4 Describe the symptoms of Alzheimer's disease. 2 marks

5 What is a post-mortem brain lesion study? 2 marks

6 How does a brain with Alzheimer's disease differ to a healthy brain? 2 marks

6 The demand for sleep

6.1 What is consciousness?

Key knowledge
- sleep as a psychological construct that is broadly categorised as a naturally occurring altered state of consciousness and is further categorised into REM and NREM sleep, and the measurement of physiological responses associated with sleep, through electroencephalography (EEG), electromyography (EMG), electro-oculography (EOG), sleep diaries and video monitoring

6.1.1 Types of sleep

Key science skills
Analyse, evaluate and communicate scientific ideas
- discuss relevant psychological information, ideas, concepts, theories and models and the connections between them

There are two types of sleep – REM and NREM. This activity will help you compare and contrast the features, physiological experiences and purposes of REM and NREM sleep.

Materials

Scissors, glue

What to do

1. Cut out the characteristics of sleep from Table 6.1.
2. Glue in the characteristics from Table 6.1 in the appropriate column of table 6.2 to indicate whether they are characteristics of NREM or REM sleep.

Table 6.1 Characteristics of NREM and REM sleep

Brainwaves range from alpha to delta waves in this type of sleep.	This type of sleep increases in duration as the night progresses.
This type of sleep is thought to be important for replenishing the mind.	During this type of sleep, sleepwalking can occur.
This type of sleep decreases in duration as the night progresses.	There are slow, rolling eye movements during this type of sleep.
Dreams commonly occur during this type of sleep.	During this type of sleep, we are in a state of paralysis.
Physiological responses, such as heart rate, increase during this type of sleep.	Our muscles are relaxed during this type of sleep.
There are rapid, erratic eye movements during this type of sleep.	This type of sleep is thought to be important for replenishing the body.
Brainwaves are beta-like waves in this type of sleep.	Physiological responses, such as heart rate, slow down during this type of sleep.

Table 6.2 Characteristics of NREM and REM sleep

NREM sleep	REM sleep

6.1.2 Brainwaves

> **Key science skills**
> Generate, collate and record data
> - organise and present data in useful and meaningful ways, including tables, bar charts and line graphs
>
> Analyse, evaluate and communicate scientific ideas
> - discuss relevant psychological information, ideas, concepts, theories and models and the connections between them

Develop

This activity will help you to identify which brainwave is which, in terms of their description, what they look like and when they are experienced.

PART A

Follow the steps to complete Table 6.3.

1. In the second column of the table, insert the name of the brainwave(s) experienced at each state/stage. Choose from the list below.
    - Theta and delta waves
    - Alpha and theta waves
    - Beta-like waves
    - Theta waves
    - Beta waves
    - Alpha waves

2. In the third column of the table, insert a description for each brainwave. Choose from the descriptions below.
    - Moderate – high amplitude, low–moderate frequency
    - Moderate amplitude, moderate frequency
    - Irregular brainwaves with a saw-tooth pattern, of low amplitude and high frequency
    - Low–moderate amplitude, moderate–high frequency
    - Low amplitude, high frequency
    - Low amplitude, high frequency with waves gradually becoming higher

3 In the fourth column of the table, draw an example of each brainwave.

Table 6.3 Brainwaves

State or stage	Name of brainwave	Description of brainwave	Illustration of brainwave
Awake			
Drowsy			
Stage 1 (NREM)			
Stage 2 (NREM)			
Stage 3 (NREM)			
REM sleep			

PART B

Label the amplitude and frequency of the brainwave shown in Figure 6.1 and provide a definition for each of these terms.

Figure 6.1 Brainwave

Amplitude:
_____
_____

Frequency:
_____
_____

6.1.3 Physiological changes and stages of sleep

> **Key science skills**
> Generate, collate and record data
> - organise and present data in useful and meaningful ways, including tables, bar charts and line graphs
>
> Analyse, evaluate and communicate scientific ideas
> - discuss relevant psychological information, ideas, concepts, theories and models and the connections between them

This activity will help you to identify which state or stage of sleep someone is experiencing by looking at measurements taken as physiological changes occur in the body. Complete this activity by answering the questions below.

1. What do these devices measure?

   a Electroencephalograph
   _____
   _____

   b Electro-oculograph
   _____
   _____

   c Electromyograph
   _____
   _____

CHAPTER 6 / The demand for sleep

2 Using your knowledge of these measures, decide whether the patients in Table 6.4 are awake, experiencing REM sleep or experiencing stage 1, 2 or 3 of NREM sleep. Provide an explanation for your decision.

Table 6.4 EEG, EOG and EMG recordings of patients in different stages of sleep and wakefulness

Patient	Physiological recordings	State or stage	Explanation
1	EEG / EOG / EMG		
2	EEG / EOG / EMG		
3	EEG / EOG / EMG		
4	EEG / EOG / EMG		
5	EEG / EOG / EMG		

6.2 Regulation of sleep–wake patterns

Key knowledge
- regulation of sleep–wake patterns by internal biological mechanisms, with reference to circadian rhythm, ultradian rhythms of REM and NREM stages 1–3, the suprachiasmatic nucleus and melatonin

6.2.1 Circadian rhythms

Key science skills
Generate, collate and record data
- organise and present data in useful and meaningful ways, including tables, bar charts and line graphs

Analyse, evaluate and communicate scientific ideas
- discuss relevant psychological information, ideas, concepts, theories and models and the connections between them

Develop

This activity will reinforce your understanding of circadian rhythms and the role they play in the human sleep–wake cycle.

PART A

Place a tick in the correct column to indicate whether each statement is true or false.

Statement	True	False
1 Circadian rhythms consist of regular automatic psychological changes that occur during a 24-hour cycle that regulate a person's chemical and hormonal production and metabolism.		
2 Ultradian rhythms are biological rhythms that follow a cycle of less than 24 hours.		
3 Circadian rhythms are influenced by a number of environmental cues, including external temperature and light levels.		
4 Light is the main environmental stimulus that adjusts our circadian rhythms to a 24-hour cycle.		
5 The sleep–wake cycle refers to the psychological pattern of alternating sleep with wakefulness over a 24-hour period.		
6 The hypothalamus, a tiny cluster of approximately 20 000 neurons located in the brain's suprachiasmatic nucleus, controls the secretion of melatonin.		
7 When external light levels are low, the pineal gland signals the suprachiasmatic nucleus to secrete more melatonin.		
8 An increase in melatonin secretion causes us to feel relaxed and sleepy. An increase in cortisol secretion causes us to feel aroused and alert.		
9 The suprachiasmatic nucleus is the brain's control centre for most of our circadian rhythms.		
10 The suprachiasmatic nucleus has direct neural links to our eyes' retinas and our brain's pineal gland.		

PART B

Answer the following questions.

1. a. What are circadian rhythms?

   b. What role do they play in the sleep–wake cycle?

2. What is the sleep–wake cycle?

3. What happens if a person experiences a shift in their sleep–wake cycle?

PART C

Revise your notes on the biological basis of the sleep–wake cycle and then complete Table 6.5. Try to do this without referring to your notes to improve your retrieval of information.

Table 6.5 Concepts and structures associated with the sleep–wake cycle

	What is it? or What are they?	Role in sleep–wake cycle
Light		
Retinas		
Optic nerve		
Suprachiasmatic nucleus (SCN)		
Pineal gland		
Melatonin		
Zeitgebers		

6.2.2 Stages of sleep

Key science skills
Generate, collate and record data
- organise and present data in useful and meaningful ways, including tables, bar charts and line graphs

Analyse, evaluate and communicate scientific ideas
- discuss relevant psychological information, ideas, concepts, theories and models and the connections between them

In this activity you will improve your knowledge of the stages of sleep.

PART A

Fill in the blanks in the sentences.
1. During a typical night's sleep, one sleep cycle lasts for approximately _____ minutes.
2. When a person first falls asleep, they exhibit slow, rolling eye movements that are typical of the _____ state.
3. In the first cycle of sleep, a person progresses from stage 1 NREM sleep through to stage _____ NREM sleep.
4. In the first cycle of sleep, a person experiences approximately _____ minutes of REM sleep.
5. As the night progresses, the periods of REM sleep _____, whereas the periods of NREM sleep _____.
6. Delta waves begin to emerge in the _____ stage of NREM sleep.
7. By the end of a night's sleep, a person can spend up to _____ minutes in REM sleep.
8. An adult will experience approximately _____ cycles of sleep per night.

PART B

Use the information from Part A to draw a line graph of the experience of a typical night's sleep. Use the axes provided in Figure 6.2.

Figure 6.2 Graph of a typical night's sleep

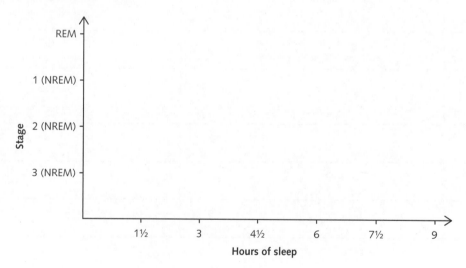

6.2.3 Evaluation of research

Key science skills
Develop aims and questions, formulate hypotheses and make predictions
- identify independent, dependent and controlled variables in controlled experiments
- formulate hypotheses to focus investigations

Analyse and evaluate data and investigation methods
- identify and analyse experimental data qualitatively, applying where appropriate concepts of: accuracy, precision, repeatability, reproducibility and validity; errors; and certainty in data, including effects of sample size on the quality of data obtained
- evaluate investigation methods and possible sources of error or uncertainty, and suggest improvements to increase validity and to reduce uncertainty

Construct evidence-based arguments and draw conclusions
- evaluate data to determine the degree to which the evidence supports the aim of the investigation, and make recommendations, as appropriate, for modifying or extending the investigation
- evaluate data to determine the degree to which the evidence supports or refutes the initial prediction or hypothesis
- use reasoning to construct scientific arguments, and to draw and justify conclusions consistent with evidence base and relevant to the question under investigation
- identify, describe and explain the limitations of conclusions, including identification of further evidence required

Analyse, evaluate and communicate scientific ideas
- discuss relevant psychological information, ideas, concepts, theories and models and the connections between them

This activity tests your understanding of circadian rhythms and the sleep–wake cycle as well as your application of research methods in psychology.

Read the research scenario below and answer the questions that follow.

Do external cues influence circadian rhythms?

Dr Julian was interested in finding out whether external cues influence circadian rhythms. He wanted to investigate the sleep–wake cycle to determine whether light and noise interrupted the normal sleep–wake cycle. To test this theory, he was intending to measure melatonin levels in participants.

He recruited 50 adult male volunteer participants, who were the first 50 people to respond to a newspaper advertisement. Each participant was asked to pull a number out of a hat. All participants who drew the number 1 out of the hat were placed in group 1 and all participants who drew the number 2 out of the hat were placed in group 2. Once they were allocated to one of the two groups, participants were given the following instructions.

Group 1 – Participants were asked to come to Dr Julian's sleep laboratory at 10 p.m. to sleep for the night. Once the participants were asleep, Dr Julian turned on lights outside the rooms where participants were sleeping so that each room was no longer in complete darkness. He also played background music for the rest of the night. Participants were woken at dawn and a saliva test was administered to test for levels of melatonin.

Group 2 – Participants were asked to come to the sleep laboratory at 10 p.m. They were to sleep in rooms there for the night. Participants in this group were able to sleep all night uninterrupted. They were woken at dawn and a saliva sample was taken.

The results of the study revealed no significant difference between the melatonin levels of participants in group 1 and group 2.

1 Write a hypothesis for this experiment.

2 Identify the independent variable(s).

3 Identify the dependent variable.

4 What experimental design was used in this study? Include one advantage and one disadvantage of using this experimental design.

5 Identify the following.
   a Experimental group:
   b Control group:

6 Describe two limitations present in this study and how they may have affected the results.

7 Provide suggestions for how this experiment could be improved if it were to be repeated.

8 Which technique was used to allocate participants to the two groups? Was this a suitable technique? Why or why not?

9 What do the results suggest? What conclusions can be made?

10 What is melatonin and why is it so important in regulating the sleep–wake cycle?

6.3 The changes in sleep over the life span

Key knowledge
- differences in, and explanations for, the demands for sleep across the life span, with reference to total amount of sleep and changes in a typical pattern of sleep (proportion of REM and NREM)

6.3.1 Investigating sleep

Key science skills
Develop aims and questions, formulate hypotheses and make predictions
- formulate hypotheses to focus investigations

Plan and conduct investigations
- design and conduct investigations; select and use methods appropriate to the investigation, including consideration of sampling technique (random and stratified) and size to achieve representativeness, and consideration of equipment and procedures, taking into account potential sources of error and uncertainty; determine the type and amount of qualitative and/or quantitative data to be generated or collated

Comply with safety and ethical guidelines
- apply relevant occupational health and safety guidelines while undertaking practical investigations

Generate, collate and record data
- systematically generate and record primary data, and collate secondary data, appropriate to the investigation
- record and summarise both qualitative and quantitative data, including use of a logbook as an authentication of generated or collated data
- organise and present data in useful and meaningful ways, including tables, bar charts and line graphs

Analyse and evaluate data and investigation methods
- identify and analyse experimental data qualitatively, applying where appropriate concepts of: accuracy, precision, repeatability, reproducibility and validity; errors; and certainty in data, including effects of sample size on the quality of data obtained
- identify outliers and contradictory or incomplete data
- evaluate investigation methods and possible sources of error or uncertainty, and suggest improvements to increase validity and to reduce uncertainty

Develop

Construct evidence-based arguments and draw conclusions
- evaluate data to determine the degree to which the evidence supports the aim of the investigation, and make recommendations, as appropriate, for modifying or extending the investigation
- evaluate data to determine the degree to which the evidence supports or refutes the initial prediction or hypothesis
- use reasoning to construct scientific arguments, and to draw and justify conclusions consistent with evidence base and relevant to the question under investigation
- identify, describe and explain the limitations of conclusions, including identification of further evidence required

Analyse, evaluate and communicate scientific ideas
- use appropriate psychological terminology, representations and conventions, including standard abbreviations, graphing conventions and units of measurement
- discuss relevant psychological information, ideas, concepts, theories and models and the connections between them

Learning and practising the key science skills is a large part of learning science. This activity will give you practice in data collation methods as well as allowing you to compare the amount of sleep that people experience across the lifespan.

1. Ask 10 people you know to write down how much sleep they get per night, over three nights. Select people of a variety of ages, from newborns to the elderly – try to include at least one and no more than two people from each age group shown in Table 6.6. (The newborns' and infants' parents can report on the infants' sleeping patterns!)
2. Collate each person's average amount of sleep over the three nights, alongside their age.
3. As a class, combine your data to answer the questions below.

Answer the following questions.

1. Write a hypothesis that this study is testing.
   _____
   _____

2. Complete Table 6.6. In the middle column, fill in the average numbers of hours that each person slept over three nights, using data obtained by the entire class. In the third column, calculate the average number of hours slept for each age group.

Table 6.6 Class data

Age group	Average number of hours slept by each person over three nights	Average number of hours slept per age group
Newborns (0–1 year)		
Infants (2–4 years)		
Children (5–12 years)		
Adolescents (13–18 years)		
Adults (19–64 years)		
Elderly (65 years and above)		

3   Graph the class findings on the axes in Figure 6.3.

Figure 6.3 Results

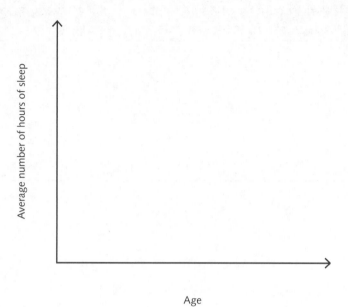

4   Was your hypothesis supported or refuted? Explain your answer.
_____
_____
_____

5   Were your findings similar to the theory discussed in class about how the amount of time spent sleeping changes over the lifespan? Explain any similarities and differences.
_____
_____
_____
_____
_____
_____

6   What could account for any differences between your findings and past research and theories?
_____
_____
_____
_____
_____

7   Why is the amount of time we spend sleeping thought to decrease with age?
_____
_____
_____

6.3.2 Review of sleep

Key science skills
Analyse, evaluate and communicate scientific ideas
- discuss relevant psychological information, ideas, concepts, theories and models and the connections between them

This activity will assist you in revising your understanding of all key knowledge areas that relate to sleep.

PART A

Use terms from Table 6.7 to fill in the blanks in the paragraphs below.

Each term is used only once.

Table 6.7 Terms related to understanding sleep

sleep–wake cycle	light	REM	pineal gland	ultradian
brain	melatonin	eyes	suprachiasmatic nucleus	NREM
90	three	dreams	active	muscles

Sleep is a naturally occurring altered state of consciousness that can be divided into two stages: REM and _____.

During _____ sleep our body is in a virtual state of paralysis. However, our _____ are rapidly moving. Our brain during this stage is highly _____ and we will experience _____.

During NREM sleep we progress through _____ distinct stages. As each stage progresses our _____ become more relaxed and the activity in our _____ starts to slow.

Deep within our brain is an internal body clock that regulates our _____. This is called the _____. This brain part directs the _____ to release a hormone that makes us feel sleepy. This hormone is called _____. This hormone also responds to the presence of _____ which is why we may feel less sleepy as morning approaches.

Throughout the night we will progress through approximately five sleep cycles. Each cycle will last for roughly _____ minutes. Each sleep cycle is an example of a(n) _____ rhythm.

PART B

1. Read the statements Table 6.8.
2. Place a tick in the correct column to indicate whether each statement is true or false.
3. If the statement is false, identify the reason why in the 'Correction' column.

Statement 1 has been completed for you to show you what to do.

Table 6.8

	True	False	Correction
1 Ultradian rhythms follow a 24-hour cycle.		✓	Ultradian rhythms follow a cycle of less than 24 hours.
2 During REM sleep, our body is restored and replenished.			
3 K-complexes occur during NREM stage 2 sleep.			
4 Children require, on average, 9 hours of sleep per night.			
5 During REM sleep our brain is highly active, with beta-like brainwaves detected on the EEG.			
6 Sleep diaries are a useful subjective measurement of sleep behaviour and other lifestyle habits that may impact sleep.			
7 Elderly people rarely experience stage 3 NREM sleep.			
8 Newborns spend roughly 50% of time in REM and 50% in NREM.			
9 Circadian rhythms follow a cycle of more than 24 hours.			
10 The suprachiasmatic nucleus is located in the cerebrum.			
11 An adult spends roughly 35% of sleep in REM.			
12 Newborns need a lot of sleep to help replenish the mental processes they exhaust when learning information during the day.			

6.3.3 Case studies

> **Key science skills**
> Analyse, evaluate and communicate scientific ideas
> - discuss relevant psychological information, ideas, concepts, theories and models and the connections between them

This activity tests your understanding of sleep and your ability to apply your knowledge to real-life situations. Read the case studies and answer the questions that follow each one.

CASE STUDY 1

Anthony's wife Larni was concerned about her husband's sleep patterns. She noticed that he often got up in the middle of the night and did strange things. One night, he attempted to make himself a sandwich; another night he tried to get into his car and drive away. Larni suggested that he spend a night in a sleep laboratory. Sleep scientists there collected data on Anthony's sleep patterns using video monitoring, EEG, EMG and EOG devices. They found that when Anthony displayed unusual sleep behaviour, his eyes were rapidly moving beneath his eyelids, his brain was highly active and his body was moving around randomly, which is uncharacteristic for this stage of sleep.

1. What condition is Anthony likely to be diagnosed with by sleep scientists?

2. What stage of sleep is Anthony in when he is engaging in the unusual behaviour? How do you know?

3. Explain why sleep scientists report that it is uncharacteristic for Anthony to be moving when in this stage of sleep.

CASE STUDY 2

Roberta has just given birth to a baby boy, Emilio. When she takes Emilio home from the hospital, she is surprised at how much time he spends asleep. All he seems to do for the first 6 weeks is sleep and eat. Roberta is looking forward to Emilio growing older, sleeping less and being a more active baby.

1. How many hours per day, on average, will Emilio sleep for the first 6 weeks of his life?

2. What percentage of his sleep will be spent in REM and NREM sleep?

3. How will Emilio's sleep patterns change as he progresses through to childhood?

CHAPTER 6 / The demand for sleep

CASE STUDY 3

Lidia and Carl are in their 80s. Their evenings are usually spent playing bingo at their local bingo club. They usually return at approximately 8 p.m. Lidia will then spend a few hours reading a novel and Carl will watch television. They then go to sleep together and wake very early in the morning.

1 How many hours per night will Lidia and Carl sleep?

2 Describe how the NREM stages of Lidia's and Carl's sleep are experienced differently in their 80s compared to when they were younger.

3 What other characteristics of their sleep may be different from when they were younger?

CASE STUDY 4

Lucy is reading her 10-year-old daughter Macey a bedtime story. Macey is enjoying the book and, at the beginning of the book, she asks questions about the story. After a few pages, Lucy notices that Macey's eyes are closed. She then feels Macey's muscles spasm rapidly for a brief moment. Lucy covers Macey with a blanket, turns off the light and lets her sleep.

1 What is the name of the muscle spasm experienced by Macey?

2 What stage of sleep is Macey in when her muscles spasm? What are some other physiological characteristics of this stage of sleep?

3 How many hours will Macey sleep and what portion of her sleep will be in REM and NREM sleep?

Exam practice

Multiple choice

Circle the response that is correct or best answers the question.

1. REM sleep is often called paradoxical sleep. The reason for this is that:
   A the brainwave patterns recorded during REM sleep are similar to those recorded when a person is awake.
   B the eyeballs move as if they are scanning a scene.
   C the skeletal muscles are limp.
   D all of the above.

2. In which stage of NREM sleep do spindles mostly occur?
   A 1
   B 2
   C 3
   D They do not occur in NREM sleep.

3. When is slow-wave sleep experienced?
   A during NREM stage 3
   B during REM
   C during NREM stage 1
   D during NREM stage 2

4. When does the first period of REM sleep occur?
   A approximately 30 minutes after falling asleep
   B approximately 60 minutes after falling asleep
   C approximately 90 minutes after falling asleep
   D approximately 120 minutes after falling asleep

Use the following information to answer Questions 5–8.
The Belling family consists of Sandra and Joe, who are both 42 years old, their 16-year-old daughter Tracey and their nine-year-old son Matt. Sandra and Tracey are often sleep deprived. This is because Sandra is a nurse who works shift work. When she is on night shift, she finds it difficult to sleep during the day. Tracey is a busy teenager completing VCE. She often finds it difficult to fall asleep at a reasonable hour.

5. How many hours of sleep does Matt need each night?
   A 8 hours
   B 9 hours
   C 10 hours
   D 15 hours

6. How many hours of sleep do Sandra and Joe need each night?
   A 6 hours
   B 8 hours
   C 9 hours
   D 10 hours

7. Tracey does not feel sleepy at 9.30 p.m. This is:
   A due to the natural sleep–wake cycle shift.
   B due to her busy social commitments.
   C because she has a lot of homework.
   D all of the above.

8 When she is on night shift, Sandra finds it difficult to sleep during the day because:
   A despite her blinds being closed, light still enters her bedroom.
   B her bed is uncomfortable.
   C she has a stressful job.
   D she has so much to do during the day.

9 What is an example of an ultradian rhythm?
   A a woman's menstrual cycle
   B sleep patterns
   C the sleep–wake cycle
   D a 9-month pregnancy

10 Where is the suprachiasmatic nucleus found?
   A pineal gland
   B pituitary gland
   C hypothalamus
   D thalamus

11 When are melatonin levels at their highest?
   A in the middle of the day
   B in the middle of the night
   C first thing in the morning
   D just before we fall asleep

12 Which of the following will not disrupt our circadian rhythms?
   A crossing time zones
   B having a head cold
   C shift work
   D damage to our suprachiasmatic nucleus

13 Biological rhythms can be influenced by:
   A internal cues only.
   B external cues only.
   C both internal and external cues.
   D neither internal nor external cues.

Short answer

1 Distinguish between circadian rhythms and ultradian rhythms. 2 marks

2 Explain why sleep diaries and video monitoring are useful techniques to measure sleep. 4 marks

3 During the night, a person cycles through two types of sleep: non-rapid eye movement (NREM) sleep and rapid eye movement (REM) sleep.

   a Describe two differences between rapid eye movement (REM) sleep and non-rapid eye movement (NREM) sleep in the sleep cycle of a healthy adult. 2 marks

   b How would REM and NREM sleep differ in the sleep cycle of a healthy adolescent compared to that of an elderly person? 2 marks

Importance of sleep in mental wellbeing 7

7.1 Partial sleep deprivation

Key knowledge
- the effects of partial sleep deprivation (inadequate sleep either in quantity or quality) on a person's affective, behavioural and cognitive functioning, and the affective and cognitive effects of one night of full sleep deprivation as a comparison to blood alcohol concentration readings of 0.05 and 0.10

7.1.1 Sleep deprivation

Key science skills
Analyse, evaluate and communicate scientific ideas
- discuss relevant psychological information, ideas, concepts, theories and models and the connections between them

Feeling tired? You could be suffering from either partial or total sleep deprivation. This activity will help you to distinguish between these two types of sleep deprivation.

PART A

Complete Figure 7.1 by providing definitions of each type of sleep deprivation.

Figure 7.1 Types of sleep deprivation

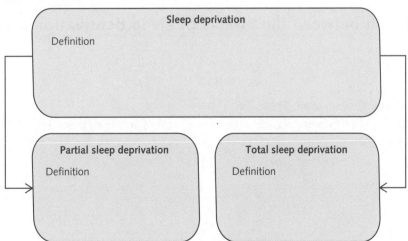

PART B

There are a number of typical effects of partial sleep deprivation. These may influence your affective, cognitive or behavioural functioning. Place a tick in the appropriate column in Table 7.1 to indicate which aspect of functioning is influenced by the listed effects of partial sleep deprivation.

Are you sleep deprived? If so, it could be affecting your functioning.

Table 7.1 Effects of sleep deprivation

Effect of sleep deprivation	Affective functioning	Cognitive functioning	Behavioural functioning
Memory lapses			
Difficulty completing routine tasks			
Difficulty maintaining attention and concentration			
Mood swings			
Difficulty processing information			
Reduced motivation			
Increase in risk-taking behaviour			
Difficulty thinking logically and problem-solving			
Increase in negative emotions			
Inability to cope with stress			
Slowed reflexes			
Reduced spatial awareness			
Irritability			
Reduced creativity			
Distorted perceptions			
Easily bored			
Lack of energy (lethargy)			
Poor decision-making			
Trembling hands			
Reduced empathy towards others			

7.1.2 A comparison between the effects of sleep deprivation and alcohol consumption

Key science skills
Analyse, evaluate and communicate scientific ideas
- discuss relevant psychological information, ideas, concepts, theories and models and the connections between them

Develop

Use the terms below to complete the sentences comparing the effects of sleep deprivation with alcohol consumption.

0.05	dangerous	deprived	metabolism
17 hours	decrease	bloodstream	moderate
24 hours	depressant	loss	slower

1. Alcohol is a _____; it slows down the messages between the brain and the body.
2. The amount of alcohol in our _____ is measured as a blood alcohol concentration (BAC).
3. Factors such as gender, body size and whether there is food in the stomach will affect alcohol tolerance and _____.
4. As the BAC approaches 0.05% the effects are such that performing tasks such as driving become _____.
5. Effects of _____ alcohol consumption on consciousness include reduced inhibitions and feeling relaxed, calmer and more confident.
6. People under the influence of alcohol may also experience a _____ of self-control, impaired mobility and coordination, and _____ reaction times.
7. Research has shown that there are similarities in the _____ in cognitive performance and changes to affective functioning when participants in studies have been _____ of sleep and comparing them to those consuming alcohol.
8. If you are fully licensed in Victoria, it is deemed safe to drive a car if you have a BAC of less than _____%.
9. Once a person has been awake for _____ this is equivalent to the effects of consuming alcohol to a BAC of 0.05%.
10. At _____ of sleep deprivation the effects are equivalent to a BAC of 0.10%.

Use each term only once.

PART B

1. Investigate the similarities in the effects on functioning of sleep deprivation and alcohol consumption.
2. Complete Table 7.2 below by listing the similar effects on affective, cognitive and behavioural functioning of sleep deprivation and alcohol consumption.

Table 7.2 Similarities between the effects of sleep deprivation and alcohol consumption

	Sleep deprivation	Alcohol consumption
Effects on affective functioning		
Effects on cognitive functioning		
Effects on behavioural functioning		

7.2 Sleep disorders

Key knowledge
- changes to a person's sleep–wake cycle that cause circadian rhythm sleep disorders (Delayed Sleep Phase Syndrome [DSPS], Advanced Sleep Phase Disorder [ASPD] and shift work) and the treatments of circadian rhythm sleep disorders through bright light therapy.

7.2.1 Circadian phase disorders and shifts in the adolescent sleep–wake cycle

Key science skills
Analyse, evaluate and communicate scientific ideas
- discuss relevant psychological information, ideas, concepts, theories and models and the connections between them

In this activity you will test your understanding of the nature of circadian phase disorders and shifts in the adolescent sleep–wake cycle. See if you can complete this activity without referring to your textbook. This will improve your retrieval of information.

PART A

Complete Figure 7.2 by providing definitions for each of the terms associated with circadian phase disorders and listing the influencing intrinsic and extrinsic factors.

Figure 7.2 Circadian phase sleep disorders

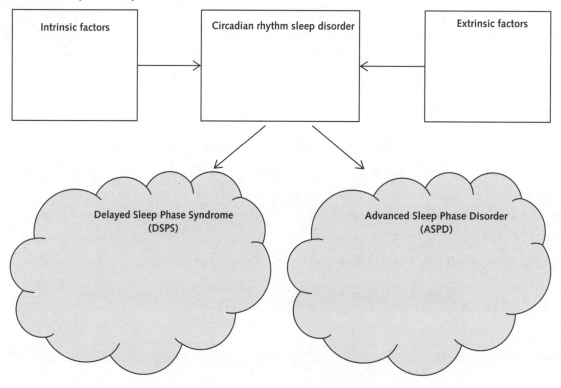

PART B

Circadian rhythms are physical, mental and behavioural changes that follow a 24-hour cycle. These rhythms regulate a number of our body processes, including chemical and hormonal production, metabolism and our sleep–wake cycle.
   Complete Figure 7.3 by inserting the labels listed below into the diagram.
- Best coordination
- Bowel movement likely
- Bowel movement suppressed
- Fastest reaction time
- Greatest cardiovascular efficiency and muscle strength
- Deepest sleep
- Highest alertness
- Highest blood pressure

- Highest body temperature
- Lowest body temperature
- Melatonin secretion starts
- Melatonin secretion stops
- Sharpest blood pressure rise

Figure 7.3 Human circadian rhythms are automatic physiological changes that regularly occur during a 24-hour cycle

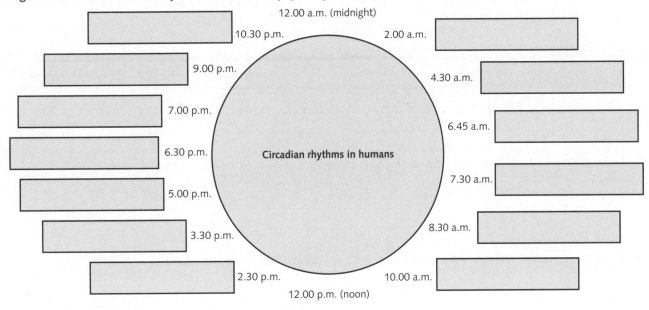

7.2.2 Sleep–wake shifts in adolescence

Key science skills
Analyse, evaluate and communicate scientific ideas
- discuss relevant psychological information, ideas, concepts, theories and models and the connections between them
- analyse and explain how models and theories are used to organise and understand observed phenomena and concepts related to psychology, identifying limitations of selected models/theories

Develop

Read the case study and answer the questions that follow.

The mother of a 16-year-old boy was tired of having to wake him up in the morning and was concerned about his sleeping habits. She had read about narcolepsy and sleep apnoea on the Internet and had convinced herself that her son had one of these disorders. She booked him into a sleep disorders clinic.

The teenager normally went to bed at midnight on school nights but did not feel sleepy for a long time afterwards. He would not fall asleep until 3.00 a.m., even if he turned off the lights and rested quietly in bed. Once he did fall asleep, he slept through the night. If allowed to, he woke on his own at noon and felt refreshed. Unfortunately, he needed to get up at 7.00 a.m. for school. His mother would have to wake him up in the morning because he would sleep through two alarms and could not get up on his own. Once out of bed, he found it was hard to get moving, and he would doze during breakfast, in the car on the way to school and during morning classes.

His weekend schedule was more erratic. He went bowling until 2.00 a.m. on Fridays, and would stay up on Saturday nights listening to music or surfing the Web. He would feel sleepy at approximately 3.00 a.m. and went straight off to sleep if he put off bedtime until then. His mother allowed him to sleep as late as he wanted on weekends, and he would get up on his own around

> lunchtime. He did not nap. When he slept until noon, he was alert for the rest of the day.
> The family turned their basement into the boy's bedroom to give him privacy. He had an active social life with a close-knit group of friends and a steady girlfriend. He was in middle secondary and was an average student. His performance at his schoolwork fell that year because he would fall asleep in class. He smoked one pack of cigarettes per day but did not drink caffeine or use recreational drugs.
>
> Source: Adapted from Steffan, M. E. (2003). Delayed sleep phase in an adolescent. *Sleep review*, September–October. http://www.sleepreviewmag.com/2003/09/delayed-sleep-phase-in-an-adolescent

1 List the symptoms that this teenager is experiencing.

___

2 What possible physical factors are contributing to the teenager's inability to fall asleep at night?

___

3 What lifestyle factors may be contributing to the teenager's inability to fall asleep at night?

___

4 What possible diagnosis could you give this teenager?

5 Provide five pieces of advice that you could give this teenager to help him establish a healthy sleep routine.

___

7.2.3 The effects of shift work on the sleep–wake cycle

Key science skills
Analyse, evaluate and communicate scientific ideas
- discuss relevant psychological information, ideas, concepts, theories and models and the connections between them

Develop

This activity will help to clarify your understanding of how shift work affects circadian rhythms and the sleep–wake cycle.

PART A

Complete Figure 7.4.

Figure 7.4 The effects of shift work on the sleep-wake cycle

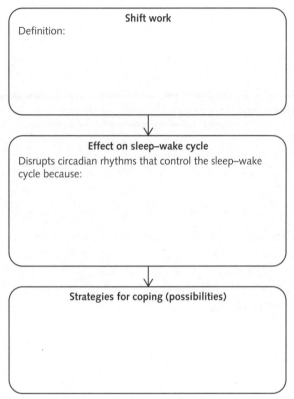

7.2.4 Bright light therapy and how to sleep better

> **Key science skills**
> Develop aims and questions, formulate hypotheses and make predictions
> - identify independent, dependent and controlled variables in controlled experiments
> - formulate hypotheses to focus investigations
>
> Construct evidence-based arguments and draw conclusions
> - use reasoning to construct scientific arguments, and to draw and justify conclusions consistent with evidence base and relevant to the question under investigation
>
> Analyse, evaluate and communicate scientific ideas
> - use appropriate psychological terminology, representations and conventions, including standard abbreviations, graphing conventions and units of measurement

Develop

This activity will strengthen your understanding of how bright light therapy can be used to treat sleep disorders. It will review an investigation into the use of bright light therapy to treat circadian rhythm disorders.

PART A

Complete the following questions on bright light therapy as a physiological treatment for circadian phase sleep disorders.

1 What are circadian phase sleep disorders?

2 What is bright light therapy?

3 What is the aim of bright light therapy?

4 Explain how bright light therapy is carried out.

PART B

Sleep researcher, Jane, investigated the use of bright light therapy to improve sleep quality in elderly patients at a Melbourne nursing home. All patients were suffering from circadian rhythm sleep disorders. Intervention consisted of altering the amount of bright light exposure prior to bedtime. A group of 10 patients who did not receive any intervention were used as a control. Improved sleep was determined by how long it took to fall asleep, number of times a patient woke during the night, quality (by self-report) and duration of sleep.

RESULTS

Table 7.3 contains the results of Jane's study.

Table 7.3 Results from sleep study

Amount of bright light exposure (min)	Mean time to fall asleep (min)	Mean no. times patient woke during night	Mean quality of sleep (scale 1–10)	Mean duration of sleep (hours)
0	19	2	4	6.3
15	15	2	5	6.7
30	12	1	7	7.1
45	10	1	8	7.5
60	7	0	10	8.2

1 Write a research hypothesis for this study.

2 Identify the independent variable.

3 Identify the dependent variable(s).

4 What is the population for this study?

5 Identify one way that Jane could have selected a sample from this population.

6 Why did 10 patients not receive any intervention?

_____

7 In Table 7.3, why is the mean shown for each measurement?

_____

8 Use the graph paper provided below to draw a graph of the mean time it took patients to fall asleep in each condition.

9 Jane removed four patient results from her data before she calculated the means for each measurement of improved sleep. Why would she do this?

_____

10 Write a conclusion for Jane's study.

_____
_____
_____
_____

7.3 Improving sleep–wake patterns and mental wellbeing

Key knowledge
- improving sleep hygiene and adaptation to zeitgebers to improve sleep–wake patterns and mental wellbeing, with reference to daylight and blue light, temperature, eating and drinking patterns

7.3.1 Improving sleep hygiene

Key science skills
Analyse, evaluate and communicate scientific ideas
- use clear, coherent and concise expression to communicate to specific audiences and for specific purposes in appropriate scientific genres, including scientific reports and posters

Use the space provided to create a pamphlet designed for adolescents, providing advice on how to improve their sleep hygiene.

Some key points to include:
- Why sleep is important
- How food and drink may influence sleep
- What is, and how to create, a bedtime ritual
- Tips on getting comfortable in bed
- How daytime naps can influence sleep quality and quantity
- The influence of stress and how to manage it for better sleep
- Where to seek additional support.

7.3.2 Analysis of a research investigation

> **Key science skills**
> Develop aims and questions, formulate hypotheses and make predictions
> - identify, research and construct aims and questions for investigation
> - identify independent, dependent and controlled variables in controlled experiments
> - formulate hypotheses to focus investigations
>
> Analyse and evaluate data and investigation methods
> - evaluate investigation methods and possible sources of error or uncertainty, and suggest improvements to increase validity and to reduce uncertainty

Develop

This activity will strengthen your research skills and reinforce your understanding of the effects of sleep deprivation on a person's affective functioning.

Read the following information and then answer the questions.

Lack of sleep tampers with your emotions: Study pinpoints neural mechanism responsible for impaired neutrality due to sleep loss

Cranky or grumpy after a long night? If so, according to a new study by Tel Aviv University, if you get less than six hours of sleep per night, your brain's ability to regulate emotions is probably compromised by fatigue. The study identifies the neurological mechanism responsible for disturbed emotion regulation and increased anxiety due to only one night's lack of sleep and reveals the changes sleep deprivation can impose on our ability to regulate emotions and allocate brain resources for cognitive processing.

'Prior to our study, it was not clear what was responsible for the emotional impairments triggered by sleep loss,' said lead researcher, Professor Talma Hendler. 'We assumed that sleep loss would intensify the processing of emotional images and thus impede brain capacity for executive functions. We were actually surprised to find that it significantly impacts the processing of both neutral and emotionally-charged images. It turns out we lose our neutrality. The ability of the brain to tell what's important is compromised. It's as if suddenly everything is important,' she said.

For the study, 18 adults were asked to take two rounds of tests while undergoing brain mapping (fMRI and/or EEG). Each round consisted of multiple tests of different types. The first round of tests was given after they had experienced a good night's sleep. The second round was given after they had been kept awake all night in a laboratory. One of the tests required participants to describe in which direction small yellow dots moved over distracting images classified as 'positively emotional' (a cat), 'negatively emotional' (a mutilated body), or 'neutral' (a spoon).

When participants had a good night's rest, they identified the direction of the dots hovering over the neutral images faster and more accurately, and their EEG pointed to differing neurological responses to neutral and emotional distractors. When sleep-deprived, however, participants performed badly in the cases of both the neutral and the emotional images, and their electrical brain responses, as measured by EEG, did not reflect a highly different response to the emotional images.

The researchers suggested that this pointed to decreased regulatory processing. They also added that it could be that sleep deprivation universally impairs judgement, but it is more likely that a lack of sleep causes neutral images to provoke an emotional response.

The researchers then conducted a second experiment testing concentration levels. Participants were shown neutral and emotional images while performing a task demanding their attention while ignoring distracting background pictures with emotional or neutral content – the depression of a key or button at certain moments – while inside an fMRI scanner. This time researchers measured activity levels in different parts of the brain as they completed the cognitive task. The team found that participants after only one night of lack of sleep were distracted by every single image (neutral and emotional), while well-rested participants were only distracted by emotional images. The effect was indicated by activity change in the amygdala, a major limbic node responsible for emotional processing in the brain.

'These results reveal that, without sleep, the mere recognition of what is an emotional and what is a neutral event is disrupted. We may experience similar emotional provocations from all incoming events, even neutral ones, and lose our ability to sort out more or less important information. This can lead to biased cognitive processing and poor judgement as well as anxiety,' said Professor Hendler.

The new findings emphasise the vital role sleep plays in maintaining good emotional balance in our life for promoting mental health. The researchers are currently examining how novel methods for sleep intervention (mostly focusing on REM sleep) may help reduce the emotional dysregulation seen in anxiety, depression and traumatic stress disorders.

Source: Adapted from American Friends of Tel Aviv University. (2015). Lack of sleep tampers with your emotions: Study pinpoints neural mechanism responsible for impaired neutrality due to sleep loss. www.sciencedaily.com/releases/2015/12/151208133618.htm (accessed 6 February, 2016).

1 What was the aim of this investigation?

2 Write a hypothesis for this investigation.

3 Identify the independent variable.

4 Identify the dependent variable.

5 Identify the research design used. State one advantage and one limitation of this type of research design.

6 What conclusions can be drawn from the results?

7 Was the data collected by the fMRI and the EEG qualitative data or quantitative data?

8 Explain what the researchers would need to do to comply with the ethical guidelines of informed consent and voluntary participation.

9 Identify a possible extraneous variable that may have affected the results. How could this be controlled in the future?

7.3.3 Zeitgebers

Key science skills
Analyse, evaluate and communicate scientific ideas
- discuss relevant psychological information, ideas, concepts, theories and models and the connections between them

Develop

This activity will consolidate your knowledge and understanding of the influence of zeitgebers on sleep–wake patterns.
Zeitgebers are cues in the environment that provide signals to our brains to do things at certain times. Complete Figure 7.5 by describing how each of the represented zeitgebers can be considered to influence sleep–wake patterns.

Figure 7.5 The influence of zeitgebers on sleep–wake patterns

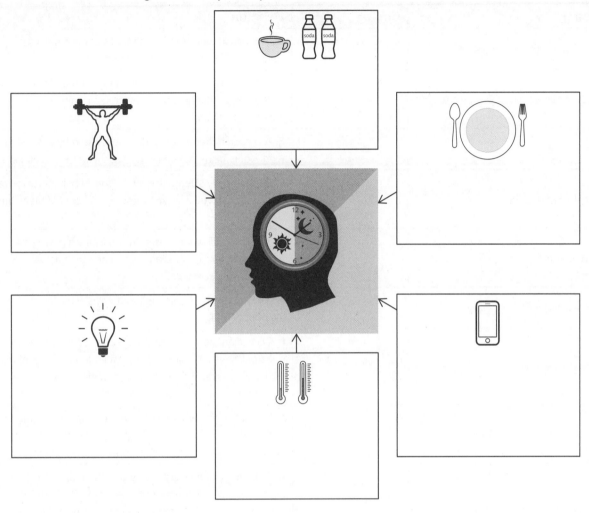

7.3.4 The importance of sleep to mental wellbeing – key terms

> **Key science skills**
> Analyse, evaluate and communicate scientific ideas
> - discuss relevant psychological information, ideas, concepts, theories and models and the connections between them

This activity will consolidate your knowledge and understanding of the key terminology for the importance of sleep to mental wellbeing.

Match each key term in Table 7.4 with its definition to review the importance of sleep to mental wellbeing. Write the letter of the definition you select in the 'Choice' column next to each term.

Table 7.4

Term	Choice	Definition
Advanced Sleep Phase Disorder (ASPD)		A. Episodes of sleep lasting only a few seconds that are not detected by the brain.
Bright light therapy		B. A natural compensatory process that occurs after being deprived of REM sleep or after periods of stress in which a person experiences increased frequency, depth and intensity of the REM stage of sleep.

Term	Choice	Definition
Circadian rhythm sleep disorder		C. The condition of not having had sufficient sleep to support optimal daytime functioning.
Delayed Sleep Phase Syndrome (DSPS)		D. An external cue such as light, temperature, noise or food that influences the activation or timing of a biological rhythm.
Microsleeps		E. A circadian rhythm sleep disorder caused by a person's work hours being scheduled during the normal sleep period.
Partial sleep deprivation		F. 24-hour daily sleep pattern that consists of approximately 16 hours of daytime wakefulness and 8 hours of night-time sleep.
REM rebound		G. Any sleep disorder caused by a mismatch between a person's internal circadian rhythm and their actual or required sleep schedule.
Shift work disorder		H. The condition of going without any sleep in a 24-hour period.
Sleep deprivation		I. A circadian rhythm sleep disorder characterised by a sleep pattern that is significantly earlier than a conventional or socially desirable sleep pattern, resulting in evening sleepiness and early-morning insomnia.
Sleep-deprivation psychosis		J. A circadian rhythm sleep disorder characterised by a sleep pattern that is significantly later than conventional sleep patterns, resulting in later sleep onset and wake times.
Sleep–wake cycle		K. A treatment for circadian rhythm phase disorders that exposes people to intense but safe amounts of artificial light for a specific and regular length of time to help synchronise their sleep–wake cycle with a normal external day–night cycle.
Total sleep deprivation		L. Getting some sleep in a 24-hour period but less than normally required for optimal daytime functioning.
Zeitgebers		M. A disruption of mental and emotional functioning as a result of lack of sleep.

Exam practice

Multiple choice

Circle the response that best answers the question.

1. Circadian rhythms are regular automatic physiological changes that
   A occur in cycles approximately every 24 hours.
   B occur in cycles approximately every 24 days.
   C occur in cycles of less than 24 hours.
   D occur in cycles of less than 24 minutes.

2. Which of the following is not controlled by circadian rhythms?
   A the sleep–wake cycle
   B the menstrual cycle
   C the release of melatonin
   D body temperature

3. Adolescents experience a shift in their sleep–wake cycle due to
   A their body releasing melatonin at an earlier time than it did previously.
   B their body releasing melatonin at a later time than it did previously.
   C excessive exposure to light emitted from electronic devices.
   D increased appetite resulting from a growth spurt.

4. Shift workers will cope best with the effects of shift work if they
   A change their shifts every week.
   B change their shift after several weeks and rotate forward one shift.
   C change their shift after several weeks and rotate backwards one shift.
   D change their shifts every two weeks.

5. Renee is extremely tired, due to lack of sleep the previous night. As a result, she is more likely to experience difficulty in performing _____ tasks, while her ability to perform _____ tasks will probably be unaffected.
   A simple; complex
   B complex; simple
   C verbal; nonverbal
   D nonverbal; verbal

6. Julianna was on an overnight school camp with her friends. After lights out, Julianna and her friends talked quietly until 2.00 a.m. After having only 4 hours sleep, the girls were woken the next morning at 6.00 a.m. Which of the following behaviours would indicate that Julianna was suffering from partial sleep deprivation?
   A She had hallucinations during the day.
   B She had decreased pain sensitivity.
   C She was very confused during the day.
   D She had difficulty focusing her eyes.

7. Microsleeps are
   A brief periods where the person appears to be awake but their brainwaves are similar to those shown in the first stage of NREM sleep.
   B brief periods where the person appears to be awake but their brainwaves are similar to those shown in REM sleep.
   C brief periods where a person appears to be asleep, but is actually awake and aware.
   D brief periods where the person shifts their attention from external to internal stimuli.

8 The aim of bright light therapy is to
   A encourage the production of melatonin.
   B discourage the production of melatonin.
   C artificially increase the length of a person's day so they feel tired when it is time for sleep.
   D synchronise a person's sleep–wake cycle with a normal external day–night cycle.

Short answer

1 Distinguish between the affective and cognitive effects of one night of full sleep deprivation in comparison to blood-alcohol concentration readings of 0.05 and 0.10%. 6 marks

2 Explain how bright light therapy may be used in the treatment of circadian rhythm sleep disorders. 2 marks

3 Describe three strategies for improving sleep hygiene. 3 marks

Defining mental wellbeing 8

8.1 Ways of considering mental wellbeing

Key knowledge
- ways of considering mental wellbeing, including levels of functioning; resilience as the ability to cope with and manage change and uncertainty; and social and emotional wellbeing (SEWB), as a multidimensional and holistic framework for wellbeing that encapsulates all elements of being (body, mind and emotions, family and kinship, community, culture, country, spirituality and ancestors) for Aboriginal and Torres Strait Islander people

8.1.1 Understanding key terms

Key science skills
Analyse, evaluate and communicate scientific ideas
- discuss relevant psychological information, ideas, concepts, theories and models and the connections between them

PART A

Table 8.1 contains some key terms associated with mental health and wellbeing. This activity has been designed to help you build your proficiency in using these key terms. Use the key terms in Table 8.1 to fill in the blanks in the following paragraphs.

Each key term is used only once.

Table 8.1 Key terms associated with ways of considering mental wellbeing

cope	poor	lifelong	challenges	positive
wellbeing	health	greater	everyday	social
low	stresses	manage	psychologically	mental
high	supportive	emotional	resilience	performance

WHAT IS MENTAL HEALTH?

The World Health Organization (WHO) defines mental _____ as 'a state of complete physical, mental and social wellbeing and not merely the absence of disease or infirmity'.

Maintaining mental health is a(n) _____ pursuit and people who attain mental health are both physically and _____ healthy.

People who enjoy a sense of wellbeing also have _____ relationships with others, do meaningful work and live in a clean environment. According to the WHO, the characteristics of a mentally healthy person include being able to realise his or her own potential, _____ with the normal stresses of life, work productively and fruitfully, and be able to make a contribution to their community. They engage in _____ thinking, show emotional resilience and are optimistic and self-confident. The typical characteristics of a mentally healthy person include _____ levels of functioning, social and emotional wellbeing and resilience to life stressors.

A key characteristic of a mentally healthy person is having a high level of functioning, which can be demonstrated by being able to interact and involve oneself in society and to undertake _____ tasks such as personal hygiene, going to work or eating food. If a person did not regularly wash or eat, or was unable to hold a job, this would show a _____ level of functioning and may indicate the presence of a mental health problem or illness.

_____ is a positive state related to the enjoyment of life, feeling connected to others, the ability to deal with challenges, and having a strong sense of purpose and control. _____ wellbeing is a sense of belonging to a community. This may involve having a job, being a member of a sporting team and making a contribution to society. _____ wellbeing means to experience positive emotions like happiness, joy or love, and feeling generally satisfied with life.

Good mental health is associated with _____ economic success, better social relationships and reduced risk of physical illness. In addition, it has a significant effect on a person's _____ in the labour market. People with _____ mental health are less attached to the labour force, work fewer hours, lose more work days and earn lower wages.

The ability of a person to adapt and cope with adversity is known as _____. A resilient person is better able to cope with life's _____, such as negative life events, and to maintain their social and emotional wellbeing. Resilience doesn't remove the _____ involved in negative events but it does enable the person to cope and better _____ stress so they are able to function more effectively and enjoy good physical and _____ health.

Try to fill in the terms without checking their definition in a textbook.

PART B

This activity has been designed to assist you in deciding whether a characteristic is mentally healthy or unhealthy. Place a tick in the correct column to indicate whether each statement would most likely be a characteristic of a mentally healthy person or a characteristic of a mentally unhealthy person.

Characteristic	Mentally healthy	Mentally unhealthy
Engages in positive thinking		
Lacks self-confidence		
Is actively involved in a variety of social and sporting groups		
Is unable to cope with the normal stresses of life		
Actively contributes to their community by volunteering to deliver food to the elderly		
Finds it difficult to move on from a disappointment		
Finds it difficult to make their own decisions		
Is unable to hold a job for an extended period of time		
Has high levels of physical health		
Enjoys challenging themselves by trying new things		

8.1.2 Resilience

> **Key science skills**
> Analyse, evaluate and communicate scientific ideas
> - discuss relevant psychological information, ideas, concepts, theories and models and the connections between them

This activity will reinforce your understanding of how resilience is important in the ability to cope with and manage change and uncertainty.

PART A

Use the terms in Table 8.2 to fill the blanks in the following paragraphs.

Each term is used only once.

Table 8.2 Terms associated with resilience as a positive adaptation to adversity

adapt	relationships	bounce back	stresses
cope	thinking	function	decrease
connectedness	resilience	challenges	restore
adversity	emotional	biology	functioning

_____ is a person's ability to successfully _____ to stress and cope with adversity, influenced by coping strategies, adaptive ways of _____ and social _____.

A resilient person is able to _____ from adversity and restore positive _____.

A resilient person is better able to cope with life's _____, such as negative life events and to maintain their social and _____ wellbeing.

Resilience doesn't remove the _____ involved in negative events but it does enable a person to _____ and better manage stress so they are able to _____ more effectively and enjoy good physical and mental health.

Resilient people can still experience a _____ in their social and emotional wellbeing during periods of _____, however, they are able to effectively resolve the situation and _____ their positive functioning.

Resilience is influenced by a person's ways of thinking, behaviours and _____, and social and situational factors. Having supportive _____ is a key factor in resilience.

PART B

Resilient people have skills and capacities that contribute to them maintaining positive social and emotional health and wellbeing. Match each of the following skills and capacities with the correct description.

Write the letter of the description you select to the right of each skill/capacity.

Skill/capacity	Choice	Description
Take care of yourself		A. Set aside time for activities, hobbies and projects you enjoy.
Be aware of what triggers your stress and how you react		B. Learn a new skill or take on a challenge to meet a goal. Learning improves your mental fitness, while striving to meet your own goals builds skills and confidence and gives you a sense of progress and achievement.
Take time to enjoy life		C. Volunteer your time for a cause or issue that you care about.
Notice the here and now		D. Be active and eat well – this helps maintain a healthy body. Combine physical activity with a balanced diet to nourish your body and mind and keep you feeling good, inside and out.

Skill/capacity	Choice	Description
Connect with others		E. You may be able to avoid some of the triggers and learn to prepare for or manage others.
Rest and refresh		F. Join a club or group of people who share your interests.
Contribute to your community		G. Develop and maintain strong relationships with people around you who will support and enrich your life.
Challenge yourself		H. Take a moment to notice each of your senses each day. Practising mindfulness, by focusing your attention on being in the moment, is a good way to do this.
Participate and share interests		I. The perfect, worry-free life does not exist. Everyone's life journey has bumpy bits and the people around you can help.
Ask for help		J. Get plenty of sleep. Go to bed at a regular time each day and practise good habits to get better sleep. Sleep restores both your mind and body.

8.1.3 Factors that contribute to positive social and emotional health and wellbeing

Key science skills
Analyse, evaluate and communicate scientific ideas
- discuss relevant psychological information, ideas, concepts, theories and models and the connections between them

There is a range of factors that contribute to maintaining positive social and emotional health and wellbeing. These factors can be categorised as either biological, psychological or social factors. Place a tick in the appropriate column to indicate which category of contributing factor each statement refers to.

See if you can do this without checking your textbook.

	Biological factor	Psychological factor	Social factor
Getting at least 8 hours of sleep each night			
Being an active member of a local basketball team			
Eating a balanced diet			
Changing unhelpful patterns of thinking			
Participating in at least 30 minutes of physical activity every day			
Talking about problems with a counsellor			
Working on strategies to develop resilience			
Reflecting on what is good in one's life			
Going on holiday with one's family			
Drinking plenty of water during the day			

CHAPTER 8 / Defining mental wellbeing

8.1.4 Social and emotional wellbeing (SEWB)

Key science skills
Analyse, evaluate and communicate scientific ideas
- discuss relevant psychological information, ideas, concepts, theories and models and the connections between them

Develop

SEWB is a holistic framework for wellbeing that encapsulates all elements of being (body, mind, family, community, culture, Land and spirituality) for Aboriginal and Torres Strait Islander peoples and communities (Figure 8.1).

Figure 8.1 A framework of SEWB

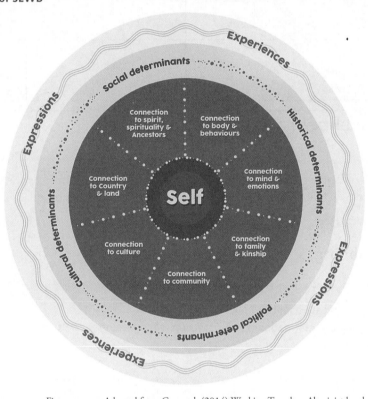

Figure source: Adapted from Gee et al. (2014) Working Together: Aboriginal and Torres Strait Islander Mental Health and Wellbeing Principles. With permission Professor Pat Dudgeon, Fact Sheet, Social and Emotional Wellbeing (https://timhwb.org.au/wp-content/uploads/2021/04/SEWB-fact-sheet.pdf).

This activity explores each of the SEWB domains with the aim of clarifying the relationship between SEWB, mental health and mental health disorders from an Aboriginal and Torres Strait Islander perspective.

Complete Table 8.3 by providing a description, risk factors and protective factors for each domain of the SEWB.

Table 8.3 The domains of SEWB with risk and protective factors

Domain	Description	Examples of risk factors	Examples of protective factors
Connection to body			
Connection to mind and emotions			
Connection to family and kinship			

Domain	Description	Examples of risk factors	Examples of protective factors
Connection to community			
Connection to culture			
Connection to *Country*			
Connection to spirituality and ancestors			

8.2 Mental wellbeing as a continuum

Key knowledge
- mental wellbeing as a continuum, with an individual's mental wellbeing influenced by the interaction of internal and external factors and fluctuating over time, as illustrated by variations for individuals experiencing stress, anxiety and phobia

8.2.1 Mental wellbeing

Key science skills
Analyse, evaluate and communicate scientific ideas
- discuss relevant psychological information, ideas, concepts, theories and models and the connections between them

Develop

This activity will help to clarify your understanding of the distinction between mental health, mental health problems and mental disorders and how mental wellbeing can be influenced by internal and external factors that can change over time.

PART A

1 Define mental health, mental health problems and mental disorders (illnesses) in the first row of Table 8.5, the mental wellbeing continuum.
2 Allocate each of the descriptive statements presented in Table 8.4 to the correct category on the continuum.

Mental wellbeing exists on a continuum and can change depending on different circumstances, which you will look at in Part B of this activity.

Table 8.4 Descriptive statements associated with mental health, mental health problems and mental disorders (illnesses)

difficulties in coping	difficulty concentrating	withdrawal and avoidance from social functions	some changes in appetite	mild to moderate stress
marked distress	able to manage feelings and emotions	able to cope with normal stressors	psychological dysfunction	physically and socially active
excessive anxiety	ongoing impairment	temporary impairment	psychological wellbeing	few sleep difficulties

Table 8.5 The mental wellbeing continuum

Mental health	Mental health problems	Mental disorders (illnesses)
Definition:	Definition:	Definition:
Descriptive statements: 1 _____ 2 _____ 3 _____ 4 _____ 5 _____	Descriptive statements: 1 _____ 2 _____ 3 _____ 4 _____ 5 _____	Descriptive statements: 1 _____ 2 _____ 3 _____ 4 _____ 5 _____

PART B

Mental health is influenced by many internal and external factors. In the table below, place a tick in the correct column to indicate whether the listed factors would be classified as either an internal or external factor that may influence a person's mental health. Complete the last four rows by listing two internal and two external factors that you have thought of yourself that may influence mental health.

Factor	Internal	External
The pressures of completing a SAC		
Someone who is confident and always seems to be able to cope		
A person who regularly experiences sufficient sleep		
A person who has recently moved schools		
Someone who has suffered a serious physical injury		
A person who has just lost their job due to being made redundant		
	✓	
	✓	
		✓
		✓

8.2.2 Biopsychosocial approach to mental wellbeing

Key science skills
Analyse, evaluate and communicate scientific ideas
- discuss relevant psychological information, ideas, concepts, theories and models and the connections between them

This activity will strengthen your understanding of how our mental wellbeing is influenced by the interaction of contributing biological, psychological and social factors.

PART A

Write each of the contributing factors listed in Table 8.6 into Figure 8.2 under either the biological, psychological or social section of the biopsychosocial model of health and illness.

Table 8.6 Contributing biological, psychological and social factors

Stress	Genetic vulnerability	Poor self-efficacy	Social support
Loss of a significant relationship	Substance use	Resilience	Poor response to medication
Poor sleep	Impaired reasoning and memory	Stigma as a barrier to accessing treatment	

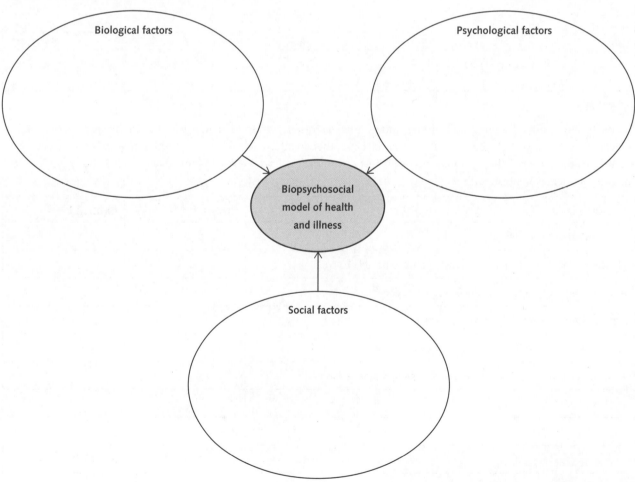

Figure 8.2 Biopsychosocial model of health and illness

PART B

Table 8.7 contains key terms associated with contributing biological, psychological and social factors. Match the terms in Table 8.7 with the definitions in Table 8.8.

Table 8.7 Contributing biological, psychological and social factors

Stress	Genetic vulnerability	Self-efficacy	Stigma
Resilience	Substance use	Social support	Sleep

Be careful with your spelling as you fill in the terms.

Table 8.8 Definitions of biological, psychological and social factors

Definition	Term
A person's belief in his or her capacity to execute behaviours necessary to produce specific performance attainments	
An increased likelihood of developing a particular disease based on a person's genetic makeup	
A person's ability to successfully adapt to stress and cope with adversity, influenced by coping strategies, adaptive ways of thinking and social connectedness	
The social disapproval of a person's personal characteristics or beliefs, or social disapproval of a type of behaviour	
A state of mental or physical tension that occurs when a person must adjust or adapt to their environment but they do not feel they have the capacity to do so	
The assistance and comfort we receive from people in our social network when we are facing a stressful or challenging situation	
The harmful or hazardous use of psychoactive substances, including alcohol and illicit drugs	
An altered state of consciousness that features the suspension of awareness of the external environment and is accompanied by a number of physiological changes to the body	

8.2.3 Stress, anxiety and phobia

Key science skills
Analyse, evaluate and communicate scientific ideas
- discuss relevant psychological information, ideas, concepts, theories and models and the connections between them

Develop

This activity will reinforce your understanding of the distinction between stress, phobia and anxiety.

PART A

Table 8.9 contains terms associated with stress, phobia and anxiety. Use these terms to fill in the blanks in the following paragraphs.

Each term is used only once.

Table 8.9 Terms associated with stress, phobia and anxiety

wellbeing	functioning	physical	anxiety	tension
phobia	perceive	excessive	cope	disorder
experienced	specific	stress	irrational	positive
exceed	fear	ongoing	challenges	subjective

_____ refers to a state of mental or physical _____ that occurs when a person must adjust or adapt to their environment, but feels they do not have the capacity to do so. Stress is experienced when the demands on a person _____ that person's perceived ability to _____. Stress is a _____ experience, which means it is a person's interpretation of the object, situation or event that determines the extent of stress experienced. Some people _____ certain events as more stressful than others. The very same situation can be _____ differently depending on the person. Although stress can be _____, such as when it helps you avoid danger or perform at your best, if it lasts for a long time, it may harm your health.

Repetitive, recurrent and _____ stress can sometimes lead to the development of a psychological disorder such as anxiety. _____ refers to feelings of apprehension, dread or uneasiness and is a response to an unclear or ambiguous threat. Anxiety is a common experience. It can be helpful in preparing for _____ that lie ahead. However, when the symptoms of anxiety include excessive and irrational worry and interfere with a person's daily _____, this may indicate the presence of a psychological disorder. When anxiety is out of proportion to a situation, it may be detrimental to a person's _____ and result in an anxiety disorder. A problem exists when intense or persistent anxiety prevents people from doing what they want or need to do.

Just as there are varying degrees of _____ health, a person's levels of anxiety can be presented on a continuum from normal feelings of being nervous to extreme feelings that can be diagnosed as symptoms of a mental disorder, such as a phobia. A _____ is a persistent, irrational and intense fear of a specific object, activity or situation. Frequently, a person can identify the factor that triggers their phobia and understand that their reaction is _____. However, they feel powerless to control their fear and may not know why it started.

A _____ phobia is an intense, irrational _____ and avoidance of a particular object (such as needles, spiders or snakes), activity (such as swimming) or situation (such as enclosed spaces). People affected by phobias recognise that their fears are unreasonable and _____, but they cannot control them. A phobia is considered to be a mental _____ because it interferes with a person's ability to function normally in everyday situations.

PART B

Table 8.10 contains the names of common phobias in the left-hand column along with descriptions of common phobias in the right-hand column. Draw lines to match the names of the common phobias with their descriptions.

Table 8.10 Common phobias

1 Acrophobia	A. The fear of falling
2 Arachnophobia	B. The fear of death
3 Cynophobia	C. The fear of dogs
4 Claustrophobia	D. The fear of spiders
5 Thanatophobia	E. The fear of being alone
6 Agoraphobia	F. The fear of driving
7 Glossophobia	G. The fear of blood
8 Aerophobia	H. The fear of chickens
9 Basiphobia	I. The fear of open or public spaces
10 Monophobia	J. The fear of commitment
11 Gamophobia	K. The fear of heights
12 Vehophobia	L. The fear of long words
13 Haemophobia	M. The fear of small spaces
14 Hippopotomonstrosesquippedaliophobia	N. The fear of public speaking
15 Alektorophobia	O. The fear of flying

Are you aware of any of these phobias?

8.2.4 Analysis of a research investigation

Key science skills

Develop aims and questions, formulate hypotheses and make predictions
- identify, research and construct aims and questions for investigation
- identify independent, dependent and controlled variables in controlled experiments
- formulate hypotheses to focus investigations

Analyse and evaluate data and investigation methods
- identify and analyse experimental data qualitatively, applying where appropriate concepts of: accuracy, precision, repeatability, reproducibility and validity; errors; and certainty in data, including effects of sample size on the quality of data obtained
- evaluate investigation methods and possible sources of error or uncertainty, and suggest improvements to increase validity and to reduce uncertainty

Construct evidence-based arguments and draw conclusions
- evaluate data to determine the degree to which the evidence supports the aim of the investigation, and make recommendations, as appropriate, for modifying or extending the investigation
- use reasoning to construct scientific arguments, and to draw and justify conclusions consistent with evidence base and relevant to the question under investigation
- identify, describe and explain the limitations of conclusions, including identification of further evidence required

Analyse, evaluate and communicate scientific ideas
- discuss relevant psychological information, ideas, concepts, theories and models and the connections between them

This activity will strengthen your research method skills and reinforce your understanding of the factors that contribute to the development and progression of mental health disorders.

Read the article below and then answer the questions that follow.

Less sleep, more struggles for elementary and middle school students

A study was conducted on 74 children from a school in the United States. The children were aged from 6 to 12. Children spent three weeks in the study. For one week, they went to bed and woke up at their normal times. For another week, they stayed up much later than normal. This meant a reduced sleeping time of 8 hours of sleep per night for children in first and second grades, and 6.5 hours for children in the third grade and up. For another week, students spent 10 hours in bed each night. To ensure compliance, subjects wore wrist monitors that logged motion.

Teachers completed a weekly survey that rated students' classroom behaviour and performance. Teachers were told that study participants would be sleeping less than usual during one of the three weeks of the study but they were not told which week that would be.

Teachers reported significantly more academic problems for students whose sleep was restricted, compared to weeks where they followed their own bedtime routine. Severity of attention problems also increased when students' sleep was restricted.

Results of the experiment complement a mounting body of evidence that inadequate sleep has an adverse effect on student academic performance.

Gahan Fallone, the study's lead author, said that we need to improve the sleeping habits of students struggling academically. For parents, he said, the message is simple: 'Getting kids to bed on time is as important as getting them to school on time'.

Source: Adapted from Brown University (2005, November 11). Less sleep, more struggles for elementary and middle school students. *ScienceDaily*. www.sciencedaily.com/releases/2005/11/051111103748.htm

1 What was the aim of this experiment?

2 What was the sample used in this experiment?

3 Identify the independent variable.

4 Identify the dependent variable.

5 Write a possible research hypothesis for this study.

6 Why weren't teachers told which week the students would have their sleep restricted?

7 State the results of the experiment.

8 How could informed consent have been obtained from the participants?

9 Identify a possible extraneous variable for this experiment and explain how it could be controlled.

8.2.5 Defining mental wellbeing concepts

Key science skills
Analyse, evaluate and communicate scientific ideas
- discuss relevant psychological information, ideas, concepts, theories and models and the connections between them

Develop

In this activity you will create study cards that will help you revise the key terms associated with defining mental wellbeing.

Materials
Poster cardboard, scissors, glue

You can use these study cards when studying for your exam.

What to do
1 Study the definitions in Table 8.11 and the terms in Table 8.12.
2 Cut around the outside of the two columns in Table 8.11. Do not cut out each cell individually at this stage.
3 Glue the columns to your poster cardboard.
4 Cut out the individual cardboard-backed rectangles of definitions. Be careful not to lose any.
5 Cut out the individual rectangles for the terms in Table 8.12.
6 Match each term to its correct cardboard-backed definition and glue them back-to-back.

Table 8.11 Definitions related to mental wellbeing

One way to refer collectively to First Nations Australians (Indigenous Australians), inclusive of the wide range of nations, cultures and languages across mainland Australia and throughout the Torres Strait.	An emotional state characterised by the anticipation of danger, dread or uneasiness as a response to an unclear or ambiguous threat.
A class of mental health disorders that feature feelings of fear, panic and the anticipation of danger, preventing normal functioning; accompanied by physical symptoms associated with threat, such as increased heart-rate, muscle tension, sweating, and rapid breathing.	An approach that proposes that health and illness outcomes are determined by the interaction and contribution of biological, psychological and social factors.
Behavioural and cognitive tactics used to manage crises, conditions and demands that are appraised as distressing.	The integration of people within their culture and the methods through which traditional knowledge is maintained and transmitted; contemporary preservation of traditional culture, including language.
The positive and protective influence of a continuing connection to traditional cultural practices to maintaining mental health, especially within First Nations (Indigenous) communities.	An increased likelihood of developing a particular disorder or condition based on a person's inherited genetic features.
A state of wellbeing in which the person realises his or her own abilities, can cope with the normal stresses of life, can work productively and fruitfully, and is able to make a contribution to his or her community.	Any condition characterised by cognitive and emotional disturbances, abnormal behaviours, impaired functioning or any combination of these, which cannot be accounted for solely by environmental circumstances.
A disruption to how a person thinks, feels and behaves, but to a lesser extent than a mental health disorder.	The experience of positive emotions like happiness, joy or love, and feeling generally satisfied with life.
A persistent, irrational and intense fear of a specific object, activity or situation.	A person's ability to successfully adapt to stress and cope with adversity, influenced by coping strategies, adaptive ways of thinking and social connectedness.
The core framework for understanding the determinants of physical and mental health for Aboriginal and Torres Strait Islander peoples; a holistic concept that results from a network of relationships between people, family, kin and community, including the importance of connection to land, culture, spirituality and ancestry, and how these affect the individual.	A kind of anxiety disorder in which a person experiences intense fear or anxiety towards a particular object or situation; the response is persistent, overwhelming and out of proportion to the actual threat.
A negative social attitude about a characteristic of a person or social group that implies some form of deficiency, often leading to unfair discrimination against or exclusion of the person or social group.	The object, entity or event that causes a feeling of stress.

9780170465069

Table 8.12 Terms related to mental wellbeing

Aboriginal and/or Torres Strait Islander peoples	Anxiety
Anxiety disorder	Biopsychosocial model
Coping strategies	Cultural continuity
Cultural determinants of mental wellbeing	Genetic vulnerability
Mental health	Mental health disorder (mental illness)
Mental health problem	Mental wellbeing
Phobia	Resilience
Social and emotional wellbeing (SEWB)	Specific phobia
Stigma	Stressor

Table 8.12 Terms related to mental wellbeing

CHAPTER 8 / Defining mental wellbeing

Exam practice

Circle the response that best answers the question.

Multiple choice

Make sure you answer every question. Never leave a multiple choice question unanswered.

1 The psychological state of someone who is functioning at a satisfactory level of emotional and behavioural adjustment is known as
   A illness.
   B wellbeing.
   C mental illness.
   D mental health problem.

2 Which one of the following is not a characteristic of a mental health problem?
   A It disrupts a person's usual level of social and emotional wellbeing.
   B It reduces a person's physical and mental abilities.
   C It lasts for a lengthy period of time.
   D None of the above.

3 The ability to bounce back during or following difficult times is known as
   A resilience.
   B spirit.
   C mental health.
   D a mental health problem.

4 Which of the following may trigger a mental health problem?
   A grief following the loss of a loved one
   B relationship difficulties
   C loss of a job
   D all of the above

5 A psychological state characterised by emotional difficulties that lead to emotional or behavioural impairment or disability serious enough to require psychiatric intervention is known as
   A a physical illness.
   B mental wellbeing.
   C mental illness.
   D a mental health problem.

6 A person suffering mild to moderate stress, temporary impairment and difficulties in coping is most likely experiencing
   A a physical illness.
   B mental wellbeing.
   C mental illness.
   D a mental health problem.

7 Which one of the following is not one of the main characteristics of a mentally healthy person?
   A a high level of functioning
   B unable to hold a job
   C able to interact and involve oneself in society
   D able to undertake everyday tasks, such as personal hygiene

8 Which one of the following would indicate a very high level of functioning?
   A mild anxiety before exams and occasional arguments with family members
   B life's problems never seem to get out of hand
   C occasional panic attacks, no friends, and unable to keep a job
   D persistent inability to maintain personal hygiene

9 Which one of the following skills and capacities are not experienced by resilient people?
   A the ability to understand and talk about one's own emotions and the feelings of others
   B the ability to form and maintain positive and respectful relationships
   C the inability to analyse a problem and make a decision
   D all of the above

10 Studies have shown that chronic stress has a significant effect on which bodily system that ultimately can trigger the onset or exacerbation of a mental health problem?
   A skeletal system
   B endocrine system
   C digestive system
   D immune system

11 Studies have found that genetic factors underpinning which neurotransmitter appear to influence the effect of stressful life events on depression?
   A dopamine
   B serotonin
   C acetylcholine
   D glutamate

Short answer

1 Distinguish between mental wellbeing and mental disorder.   4 marks

2 What is the SEWB framework?   2 marks

3 Using an example, explain why a specific phobia is considered to be a mental health disorder.   2 marks

Application of a biopsychosocial approach to explain specific phobia

9

9.1 Development of specific phobia

Key knowledge
- the relative influences of factors that contribute to the development of specific phobia, with reference to gamma-aminobutyric acid (GABA) dysfunction and long-term potentiation (biological); behavioural models involving precipitation by classical conditioning and perpetuation by operant conditioning, and cognitive biases including memory bias and catastrophic thinking (psychological); and specific environmental triggers and stigma around seeking treatment (social)

9.1.1 Factors that contribute to the development of specific phobia

Key science skills
Analyse, evaluate and communicate scientific ideas
- discuss relevant psychological information, ideas, concepts, theories and models and the connections between them

Develop

PART A

Table 9.1 contains terms associated with factors that contribute to the development of a specific phobia. Use the terms in Table 9.1 to fill in the blanks in the following sentences.

Each term is used only once.

Table 9.1 Terms associated with factors that contribute to the development of specific phobia

amygdala	biological	social learning theory
discrimination	flight-or-fight-or-freeze response	stigma
low	GABA	strengthening
psychological	observing	reasoning
social factors	rationale	triggers
stimuli	biopsychosocial	predispose
vulnerable		

1 The _____ model proposes that health and illness outcomes are determined by the interaction and contribution of _____, _____ and _____.

2 Reduced _____ is related to dysfunction in the brain circuits that regulate emotional responses to threatening stimuli.

3 An individual with _____ levels of GABA may not be able to regulate their _____ and therefore may be more _____ to anxiety and have a greater chance of developing a specific phobia.

4 Long-term potentiation contributes to the development of specific phobia via the _____ of synapses within neural circuits that communicate with the _____. This activates multiple brain regions that then produce a variety of symptoms of fear and anxiety.

5 The cognitive model claims we may sometimes couple faulty _____ and _____ with fearful _____ from the environment and hence a new cognition is formed into a phobia.

6 Specific environmental _____ in the environment can _____ an individual to the development of specific phobias, such as a traumatic event.

7 According to _____ and parental modelling, children learn from their parents and often mirror what the parents do. Children learn how to respond to stress and difficulties by _____ how parents respond to these.

8 _____ is a negative social attitude about a characteristic of a person or social group that implies some form of deficiency, often leading to unfair _____ against or exclusion of the person or social group.

9.1.2 The biopsychosocial framework: application to understanding specific phobia

Key science skills
Analyse, evaluate and communicate scientific ideas
- discuss relevant psychological information, ideas, concepts, theories and models and the connections between them
- analyse and explain how models and theories are used to organise and understand observed phenomena and concepts related to psychology, identifying limitations of selected models/theories

Develop

1 Refer to your textbook or use the Internet to choose a specific phobia that you would like to know more about.
2 Research the specific phobia using your chosen resource(s).
3 Complete Table 9.2 with the information that you gather during your research.

Table 9.2 Investigating a specific phobia

Specific phobia			
Symptoms	Possible biological influences	Possible psychological influences	Possible social influences

PART B

1 Use the space provided to write a scenario involving the specific phobia you researched in Part A of this activity. Use a fictitious person and a made-up situation.
2 Fill in Figure 9.1 to explain how each of the biological, psychological and social influences may contribute to the specific phobia you investigated.

SCENARIO

Figure 9.1 Biopsychosocial model of a specific phobia

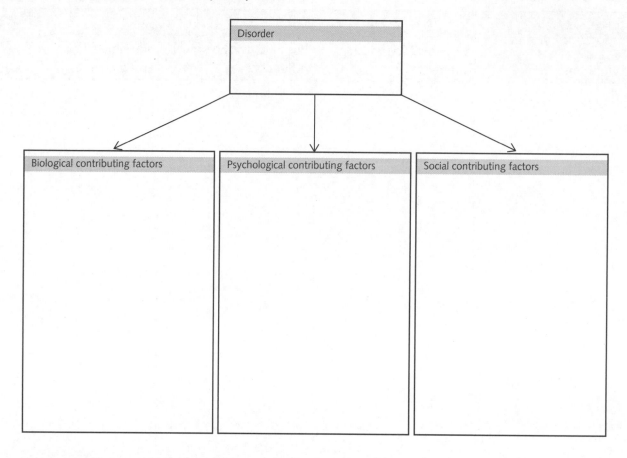

9.1.3 Development of specific phobia

Key science skills
Analyse, evaluate and communicate scientific ideas
- discuss relevant psychological information, ideas, concepts, theories and models and the connections between them

Read the following scenario and answer the questions that follow.

SCENARIO

Ann-Marie, who is 16, was referred to a disorder clinic after increasing episodes of fainting. Her friends and family were concerned about her and she had been suspended from school.

About two years earlier, in her life science class, the teacher had shown a film of a woman giving birth, to illustrate various points about anatomy and the birth process. This was a particularly graphic film, showing vivid images of blood and tissue. About halfway through, Ann-Marie felt a bit light-headed and left the room. But the images did not leave her. She continued to be bothered by them and occasionally felt slightly queasy. She began to avoid situations where she might see blood or have an accident that might expose her to blood and tissue. She found it difficult to look at raw meat, or even band-aids, because they brought the feared images to mind.

Ann-Marie quit her part-time job in the local supermarket because she couldn't bear to handle meat items. Eventually, anything her friends or parents said that evoked an image of blood or injury caused her to feel light-headed. It got so bad that if a teacher asked a student in class to 'cut it out', she felt faint.

About six months before her visit to the clinic, Ann-Marie fainted when she came into contact with a towel covered in blood after her mother accidentally cut herself while cooking. This marked the beginning of her fainting episodes.

CHAPTER 9 / Application of a biopsychosocial approach to explain specific phobia

Ann-Marie's GP could find nothing wrong with her, nor could several other physicians. At this stage, she was fainting 6 to 10 times a week. This was obviously hard for her and for other students in her class. Each time she fainted in class, the class would stop. The school principal finally concluded that she was being manipulative and that her fainting was merely a case of trying to draw attention to herself. As a result, he suspended her from school.

However, after attending the disorder clinic, Ann-Marie was found to be suffering from a phobia, which caused her physical responses.

1 Define the term 'phobia'.

2 Ann-Marie developed a phobia after watching a film that showed graphic scenes of blood. Explain why Ann-Marie was the only person in her class to develop a phobia, even though all the students saw the same film.

3 The psychologist treating Ann-Marie has decided to use systematic desensitisation. Describe how this method could be used to treat Ann-Marie's disorder.

PART B

1 Re-read the scenario in Part A, then study Figure 9.2, which shows the symptoms experienced by Ann-Marie and provides additional information about her life.
2 Using this information, label the biological, psychological and social factors in Figure 9.2 that may have contributed to Ann-Marie's phobia, and explain how these combine to present a biopsychosocial explanation for the development of her phobia.

Figure 9.2 Ann-Marie's phobia

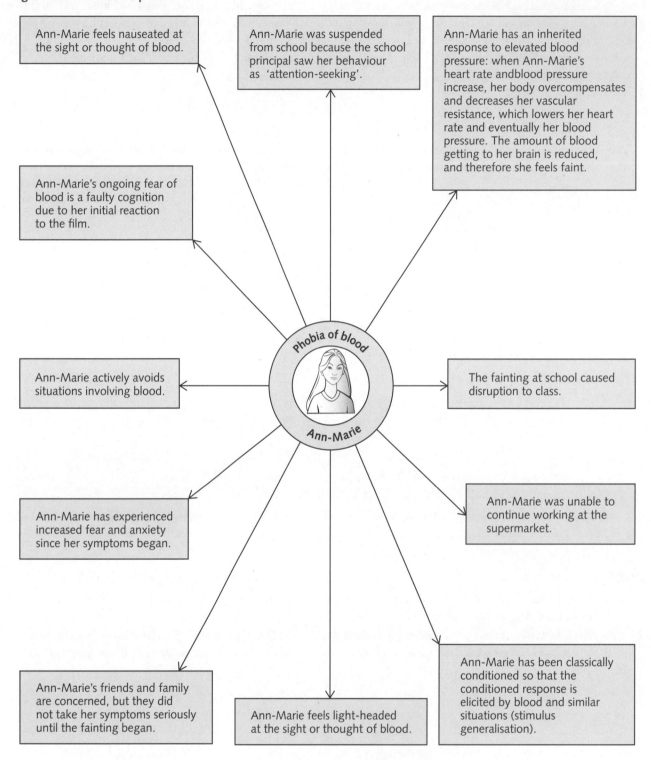

9.1.4 Behavioural factors influencing the development of specific phobia

Key science skills
Analyse, evaluate and communicate scientific ideas
- discuss relevant psychological information, ideas, concepts, theories and models and the connections between them
- analyse and explain how models and theories are used to organise and understand observed phenomena and concepts related to psychology, identifying limitations of selected models/theories

CHAPTER 9 / Application of a biopsychosocial approach to explain specific phobia 171

This activity will consolidate your understanding of the behavioural factors that contribute to the development of specific phobia. Use your own examples to complete Tables 9.3 and 9.4 to outline the influence of classical conditioning and operant conditioning on the development of a specific phobia.

PART A

Table 9.3 Precipitation of a specific phobia by classical conditioning

Before conditioning	During conditioning (acquisition)	Test for conditioning

PART B

Table 9.4 Perpetuation of a specific phobia by operant conditioning

Antecedent stimulus		Behaviour		Consequence		Behaviour maintained
	→		→		→	

9.2 Evidence-based interventions

Key knowledge
- evidence-based interventions and their use for specific phobia, with reference to the use of short-acting anti-anxiety benzodiazepine agents (GABA agonists) in the management of phobic anxiety and breathing retraining (biological); the use of cognitive behavioural therapy (CBT) and systematic desensitisation as psychotherapeutic treatments of phobia (psychological); and psychoeducation for families/supporters with reference to challenging unrealistic or anxious thoughts and not encouraging avoidance behaviours (social)

9.2.1 Terms used in evidence-based interventions

Key science skills
Analyse, evaluate and communicate scientific ideas
- discuss relevant psychological information, ideas, concepts, theories and models and the connections between them

Develop

PART A

Table 9.5 provides terms associated with evidence-based interventions. Use these terms to fill in the blanks in the following sentences.

Each term is used only once.

Table 9.5 Terms associated with evidence-based interventions

avoidance	benzodiazepines	calming
cause	cognitive behavioural therapy (CBT)	control
counterconditioning	desensitisation	experiences
information	negative	patterns
prevent	psychoeducation	psychological
relaxation	retraining	symptoms
understand	unproductive	

1. Medications can relieve _____ of anxiety; however, their use is not a long-term solution because they do not treat the underlying _____ of the anxiety disorder.
2. _____ are medications that act to promote GABA (the primary inhibitory neurotransmitter), thus _____ physiological arousal and reducing extreme anxiety associated with a specific phobia.
3. _____ techniques such as breathing _____ can help individuals cope effectively with the stresses related to their specific phobia.
4. By changing breathing _____, breathing retraining can help to correct breathing habits and help individuals have more _____ of their anxiety.
5. _____ is an evidence-based _____ treatment approach that teaches clients to apply cognitive behavioural strategies to recognise and change _____ and _____ patterns of thinking and behaving.
6. Systematic _____ is a type of behaviour therapy that uses _____ to reduce the anxiety a person _____ when in the presence of, or thinking about, a feared stimulus.
7. _____ is a psychosocial approach in which a person experiencing a mental health problem or disorder and their family are provided with _____ to help them _____ the condition and how they can contribute to managing it.
8. _____ behaviours are behaviours that attempt to _____ exposure to the fear-provoking object, activity or situation.

9.2.2 Understanding evidence-based interventions

> **Key science skills**
> Analyse, evaluate and communicate scientific ideas
> - discuss relevant psychological information, ideas, concepts, theories and models and the connections between them
> - analyse and explain how models and theories are used to organise and understand observed phenomena and concepts related to psychology, identifying limitations of selected models/theories

Develop

1. Figure 9.3 provides a template to assist you in understanding the evidence-based interventions available to treat specific phobias. Start by completing the top box by explaining what is meant by an evidence-based intervention.
2. Complete the remaining boxes by identifying and describing one biological, one psychological and one social evidence-based intervention that can be used in the treatment of a specific phobia.

Figure 9.3 Evidence-based interventions and their use for specific phobia

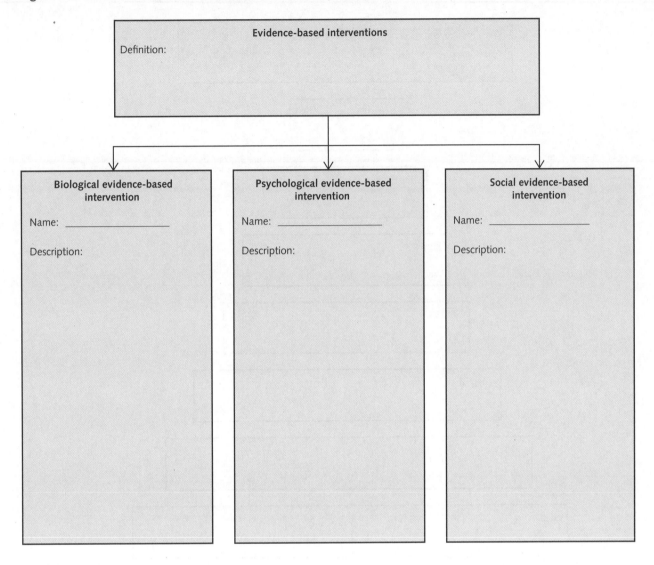

9.2.3 Systematic desensitisation

Key science skills
Analyse, evaluate and communicate scientific ideas
- discuss relevant psychological information, ideas, concepts, theories and models and the connections between them

1 Refer to your textbook or use the Internet to choose a specific phobia that you would like to know more about. You may reuse the one that you used in Activity 9.1.2.
2 Figure 9.4 provides a template for systematic desensitisation. Complete this figure by creating a fear hierarchy that could be used by a practitioner in treating someone suffering from the chosen specific phobia.

Figure 9.4 Fear hierarchy for specific phobia

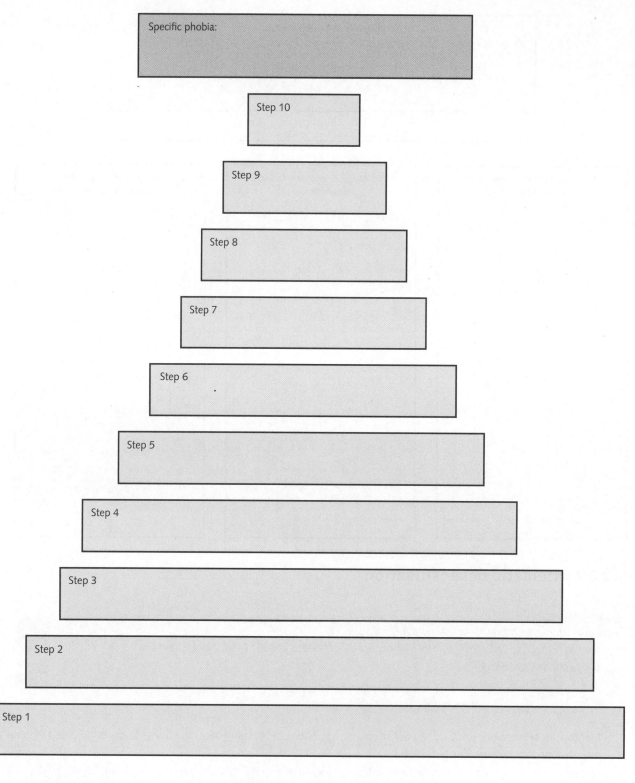

9.2.4 Biopsychosocial approach study cards

Key science skills
Analyse, evaluate and communicate scientific ideas
- discuss relevant psychological information, ideas, concepts, theories and models and the connections between them

In this activity you will create study cards that will help you revise the key terms associated with defining mental wellbeing.

Materials

Poster cardboard, scissors, glue

What to do

1. Cut around the outside of the two columns of definitions in Table 9.6. Do not cut out each cell individually at this stage.
2. Glue the columns to your poster cardboard.
3. Now cut out the individual cardboard-backed cells of definitions. Place them in a pile and be careful not to lose them.
4. Cut out the individual cells of terms in Table 9.7. Place them in a pile and be careful not to lose them.
5. Match your cut-out definitions with your cut-out terms.
6. When they are matched, glue the paper terms to the backs of the definitions already attached to the poster cardboard.
7. Use the study cards when revising this topic.

Table 9.6 Definitions related to specific phobia

In neural communication, a substance that binds to a neuroreceptor to produce a similar effect to that of a neurotransmitter in either exciting or inhibiting a postsynaptic neuron.	In neural communication, a substance that suppresses the release of a neurotransmitter or blocks the receptor sites, making the postsynaptic neuron less likely to fire.
Behaviours that attempt to prevent exposure to a fear-provoking object, activity or situation.	In specific phobia, an explanatory framework that proposes phobic anxiety is the result of learned associations.
A group of medications used in the short-term treatment of anxiety; as GABA agonists, they enhance the GABA-induced inhibition of overexcited neurotransmitters, calming nervous activity.	Medication to treat high blood pressure, congestive heart failure, abnormal heart rhythms and anxiety by reducing heart rate and blood pressure through inhibiting the binding of adrenaline (epinephrine) to neuroreceptors.
An approach that proposes that health and illness outcomes are determined by the interaction and contribution of biological, psychological and social factors.	The process of identifying incorrect breathing habits and replacing them with correct ones.
Occurs when a person repeatedly overestimates the potential dangers and assumes the worst of an object or event.	An evidence-based psychological treatment approach that teaches clients to apply cognitive behavioural strategies to recognise and change negative and unproductive patterns of thinking and behaving.
An automatic tendency or preference for processing or interpreting information in a particular way, producing systematic errors in thinking when making judgements or decisions.	A psychological approach based on understanding how people's thought patterns, memories and beliefs affect their emotions, attitudes and behaviours.

Psychological treatments whose effectiveness have been supported by the integration of clinical research findings and clinical expertise.	The primary inhibitory neurotransmitter; its overall effects are to calm or slow neural transmission and therefore the body's response.
A form of synaptic plasticity that results in a long-lasting strengthening of neural connections at the synapse as a result of repeated stimulations from a presynaptic to postsynaptic neuron.	Behaviour that is potentially harmful and prevents a person from meeting and adapting to the demands of everyday living.
A tendency to remember information of one kind at the expense of another kind; including the bias towards remembering negative and threat-related experiences that is associated with specific phobia.	The social and cognitive processes of learning by observing another person's behaviour and the consequences of the behaviour.
A psychosocial approach in which a person experiencing a mental health problem or disorder and their family are provided with information to help them understand the condition and how they can contribute to managing it.	Any psychological technique used for treating mental health disorders, with the goal of producing positive changes in thinking, emotions, personality, behaviour or adjustment.
The concept that one emotional state is used to block another, as is the case in systematic desensitisation.	A theory in which human learning is proposed to occur primarily through interactions with others, especially through observing the behaviours of others and the consequences that follow.
A negative social attitude about a characteristic of a person or social group that implies some form of deficiency, often leading to unfair discrimination against or exclusion of the person or social group.	A type of behaviour therapy that uses counterconditioning to reduce the anxiety a person experiences when in the presence of, or thinking about, a feared stimulus.

CHAPTER 9 / Application of a biopsychosocial approach to explain specific phobia

Table 9.7 Terms related to specific phobia

Agonist	Antagonist
Avoidance behaviours	Behavioural model
Benzodiazepines	Beta blockers
Biopsychosocial model	Breathing retraining
Catastrophic thinking	Cognitive behavioural therapy (CBT)
Cognitive bias	Cognitive model
Evidence-based interventions	Gamma-aminobutyric acid (GABA)
Long-term potentiation (LTP)	Maladaptive behaviour
Memory bias	Modelling
Psychoeducation	Psychotherapy
Reciprocal inhibition	Social learning theory
Stigma	Systematic desensitisation

9.2.5 Analysis of a research investigation

Key science skills

Develop aims and questions, formulate hypotheses and make predictions
- identify, research and construct aims and questions for investigation

Analyse, evaluate and communicate scientific ideas
- discuss relevant psychological information, ideas, concepts, theories and models and the connections between them
- analyse and explain how models and theories are used to organise and understand observed phenomena and concepts related to psychology, identifying limitations of selected models/theories
- analyse and evaluate psychological issues using relevant ethical concepts and guidelines, including the influence of social, economic, legal and political factors relevant to the selected issue

Read the following article and then answer the questions that follow.

Touching tarantulas: Overcoming phobias with brief therapy

Fear of spiders is a subtype of an anxiety disorder called specific phobia, one of the most common anxiety disorders affecting about seven per cent of the population. Common specific phobias include fear of blood, needles, snakes, flying and enclosed spaces.

This is the first study to document the immediate and long-term brain changes after treatment and to illustrate how the brain undergoes a reorganisation to reduce fear as a result of the therapy. The findings show the lasting effectiveness of short exposure therapy for a phobia and offer new directions for treating other phobias and anxiety disorders.

Katherina Hauner, one of the researchers, said that before treatment, some of the participants wouldn't walk on grass for fear of spiders or would stay out of their home for days if they thought a spider was present.

The therapy involved gradually approaching the spider. Before the session, the participants (12 adults) were even afraid to look at photos of spiders. When they did, the regions of the brain associated with fear response – the amygdala, insula and cingulate cortex – lit up with activity in an fMRI scan. When asked to touch a tarantula in a closed terrarium or approach it as closely as possible, they were not able to get closer than about 3 metres on average.

During the therapy, participants were taught about tarantulas and learned that their catastrophic thoughts about them were not true. Some thought that the tarantula might be capable of jumping out of the cage and on to them. Some thought the tarantula was capable of planning something evil to purposefully hurt them. Participants were taught that the tarantula is fragile and more interested in trying to hide itself, and that its movements are predictable. They gradually learned to approach the tarantula in slow steps until they were able to touch the outside of the terrarium. Then they touched the tarantula with a paintbrush, a glove and eventually with their bare hands. Some held it.

Immediately after the therapy, an fMRI scan showed that the brain regions associated with fear decreased in activity when people were again shown the spider photos, a reduction that persisted six months after treatment.

When the same participants were asked to touch the tarantula six months later, they did so without hesitation.

The researchers were also able to predict who the therapy would be most effective for by studying the person's brain activity immediately after the treatment. Participants with higher measurements of activity in brain regions associated with visual perception of fearful stimuli immediately after the treatment were much more likely to show the lowest fear of spiders six months later.

Source: Adapted from Northwestern University. "Touching tarantulas: Overcoming phobias with brief therapy." ScienceDaily, 21 May 2012

---

1 What was the aim of this study?

2 Who were the participants used in this study?

3 What type of sampling was used in this study?

4. Define the term 'phobia'. What participant behaviour described in the research indicates that the participants were suffering from a spider phobia?

5. Define CBT and systematic desensitisation and describe how each were used to treat the participants' spider phobia.

6. Identify the regions of the brain that are associated with the fear response.

7. State the results of the study.

8. How could informed consent have been obtained from the participants?

CHAPTER 9 / Application of a biopsychosocial approach to explain specific phobia

Exam practice

Multiple choice

Circle the response that best answers the question.

1 Neurotransmitters carry information about stimuli from the _____ to the _____ and affect the activity level of the receiving neuron.
   A deactivating neuron; activating neuron
   B postsynaptic neuron; presynaptic neuron
   C activating neuron; deactivating neuron
   D presynaptic neuron; postsynaptic neuron

2 Which neurotransmitter is known to influence anxiety disorders by having an inhibiting effect?
   A epinephrine
   B gamma-aminobutyric acid
   C dopamine
   D norepinephrine

3 When a person learns to fear a particular stimulus the initial learning will form a new memory circuit with established connections within the amygdala. Structural change to neural circuits is known as
   A synaptogenesis.
   B the consolidation theory.
   C long-term potentiation.
   D proliferation.

4 The strengthening of synapses based on recent patterns of activity is known as
   A synaptogenesis.
   B the consolidation theory.
   C long-term potentiation.
   D proliferation.

5 It is possible to explain phobia as a learned association between two stimuli. This is an example of
   A observational learning.
   B modelling.
   C operant conditioning.
   D classical conditioning.

6 The perpetuation of a phobia can be explained by the consequences of reward and punishment. This is an example of
   A observational learning.
   B modelling.
   C operant conditioning.
   D classical conditioning.

7 The _____ model examines the influence that inaccurate mental processes have in the development and maintenance of phobias.
   A cognitive
   B behavioural
   C cultural
   D neurological

8 When a person repeatedly overestimates the potential dangers of an object or event and assumes the worst-case scenario, it is known as
   A cognitive bias.
   B memory bias.
   C catastrophic thinking.
   D rational thinking.

9 If a child learns to fear dogs because they have seen their parent responding to a dog with fear, it can be said that this learning has occurred through the process of
   A modelling.
   B parental learning.
   C classical conditioning.
   D fear conditioning.

10 People who experience mental illness such as a specific phobia are often faced with stigma that results from a lack of understanding about their illness. Which of the following are harmful effects of stigma associated with anxiety disorders?
    A physical violence or harassment
    B fewer opportunities for work
    C reluctance to seek help or treatment
    D all of the above

11 The short-acting anti-anxiety benzodiazepine stimulates the GABA neurotransmitters that
    A increase psychological arousal.
    B reduce psychological arousal.
    C reduce physiological arousal.
    D increase physiological arousal.

12 An anxious person's breathing may consist of small, shallow breaths. This may _____ oxygen levels and _____ the amount of carbon dioxide in the blood.
    A raise; reduce
    B raise; raise
    C reduce; reduce
    D reduce; raise

13 Cognitive behavioural strategies used by health professionals will be personalised to a patient's individual needs. These strategies may include
    A teaching relaxation and breathing techniques to manage stress.
    B teaching to recognise the difference between unhelpful and productive thoughts.
    C educating patients about the body's natural reactions to threatening objects.
    D all of the above.

Short answer

1 Describe how GABA dysfunction may contribute to phobia. 3 marks

2 Explain the process of using systematic desensitisation as a treatment of phobia. 4 marks

3 Outline two goals of psychoeducation (one for the phobic person and one for the family/supporter of the phobic person). 2 marks

10 Maintenance of mental wellbeing

10.1 The application of a biopsychosocial approach to maintaining mental wellbeing

Key knowledge
- the application of a biopsychosocial approach to maintaining mental wellbeing, with reference to protective factors including adequate nutritional intake and hydration and sleep (biological), cognitive behavioural strategies and mindfulness meditation (psychological) and support from family, friends and community that is authentic and energising (social)

10.1.1 The application of a biopsychosocial approach to maintaining mental wellbeing

Key science skills
Analyse, evaluate and communicate scientific ideas
- discuss relevant psychological information, ideas, concepts, theories and models and the connections between them

Table 10.1 contains terms that are associated with ways of considering mental wellbeing. Use the terms to fill in the blanks in the following sentences.

Table 10.1 Terms associated with ways of considering mental wellbeing

maintaining	biological	adequate sleep	social support
chronic	functioning	stress	protective factor
attention	judgement	risk factors	challenging
nutritional intake	resilience	cognitive behavioural strategies	depression

1 _____ is a normal part of life and it can motivate us to achieve our goals. However, if stress becomes _____, or if we feel we're no longer able to cope, it can negatively affect our wellbeing.

2 A number of factors, known as _____, increase the likelihood that an individual will experience mental health problems or disorders.

3   Conversely, there are also many factors that influence an individual's _____ and ability to recover from a negative experience. Any behavioural, _____, psychological or environmental characteristic that decreases the likelihood of a person developing a particular mental health problem or disorder is known as a _____.

4   Brain _____ is very sensitive to what we eat and drink, therefore diet can play a vital role in _____ mental wellbeing.

5   _____ and hydration contribute to resilience by protecting against diet-related diseases that affect physical and cognitive functions and by reducing vulnerability to stress and _____.

6   _____ is the necessary amount for individuals to function effectively during the daytime and cope with normal daily stress.

7   _____ are structured psychological treatments that recognise that a person's way of thinking (cognition) and acting (behaviour) affect the way they feel.

8   Mindfulness meditation is a meditation practice in which a person focuses _____ on their breathing, with thoughts, feelings and sensations being experienced freely as they arise and without _____.

9   _____ is the assistance and comfort we receive from people in our social network when we are facing a stressful or _____ situation, from family members and friends through to support groups and social institutions.

10.1.2 Australian Guide to Healthy Eating

Key science skills
Generate, collate and record data
- organise and present data in useful and meaningful ways, including tables, bar charts and line graphs

Analyse, evaluate and communicate scientific ideas
- discuss relevant psychological information, ideas, concepts, theories and models and the connections between them

PART A

The Australian Guide to Healthy Eating provides advice about the amount and kinds of foods you need to eat to maintain health and good mental wellbeing. Figure 10.1 provides a template for the Australian Guide to Healthy Eating for you to complete by providing three examples of foods for each food group.

Figure 10.1 Template for the Australian Guide to Healthy Eating

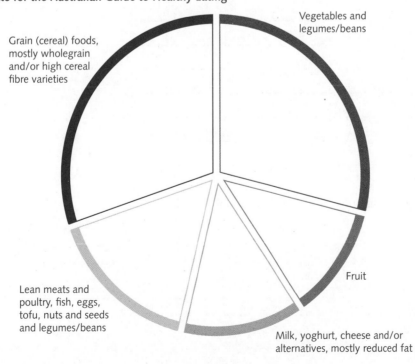

PART B

1 State and explain five ways that the choice of foods may influence mental wellbeing.

_____
_____
_____
_____
_____
_____
_____
_____
_____
_____

10.1.3 Adequate sleep

Key science skills
Generate, collate and record data
- organise and present data in useful and meaningful ways, including tables, bar charts and line graphs

Analyse, evaluate and communicate scientific ideas
- discuss relevant psychological information, ideas, concepts, theories and models and the connections between them

Develop

PART A

Adequate sleep is the necessary amount of sleep for individuals to function effectively during the daytime and cope with normal daily stress. The amount of adequate sleep will differ between individuals depending on a range of factors such as age, physical activity levels and general health. Complete Table 10.2 below by indicating the recommended hours of sleep for each age group. The first age group has been completed as a guide. See if you can complete the table without referring to your textbook.

Table 10.2 Recommended hours of sleep by age

Age group	Recommended hours of sleep
Newborn (0–3 months)	14–17
Infant (4–11 months)	
Toddler (1–2 years)	
Preschool (3–5 years)	
School age (6–13 years)	
Teenager (14–17 years)	
Young adult (18–25 years)	
Adult (26–64 years)	
Older adult (65+ years)	

PART B

1 Describe five strategies to improve sleep for better mental wellbeing.

_____
_____
_____
_____
_____

10.1.4 Analysis of a research investigation

Key science skills
Develop aims and questions, formulate hypotheses and make predictions
- identify, research and construct aims and questions for investigation
- identify independent, dependent and controlled variables in controlled experiments
- formulate hypotheses to focus investigations

Analyse and evaluate data and investigation methods
- evaluate investigation methods and possible sources of error or uncertainty, and suggest improvements to increase validity and to reduce uncertainty

Analyse, evaluate and communicate scientific ideas
- discuss relevant psychological information, ideas, concepts, theories and models and the connections between them

Read the article about mindfulness meditation and anxiety carefully and answer the questions that follow.

Just 10 minutes of meditation helps anxious people have better focus

Just 10 minutes of daily mindful meditation can help prevent your mind from wandering and is particularly effective if you tend to have repetitive, anxious thoughts, according to a study from the University of Waterloo.

The study, which assessed the impact of meditation with 82 participants who experience anxiety, found that developing an awareness of the present moment reduced incidents of repetitive, off-task thinking, a hallmark of anxiety.

'Our results indicate that mindfulness training may have protective effects on mind wandering for anxious individuals,' said Mengran Xu, a researcher and PhD candidate at Waterloo. 'We also found that meditation practice appears to help anxious people to shift their attention from their own internal worries to the present-moment external world, which enables better focus on a task at hand.'

The term mindfulness is commonly defined as paying attention on purpose, in the present moment, and without judgement.

As part of the study, participants were asked to perform a task on a computer while experiencing interruptions to gauge their ability to stay focused on the task. Researchers then put the participants into two groups at random, with the control group given an audio story to listen to and the other group asked to engage in a short meditation exercise prior to being reassessed.

'Mind wandering accounts for nearly half of any person's daily stream of consciousness,' said Xu. 'For people with anxiety, repetitive off-task thoughts can negatively affect their ability to learn, to complete tasks or even function safely.

'It would be interesting to see what the impacts would be if mindful meditation was practiced by anxious populations more widely.'

Source: Adapted from University of Waterloo. (2017, May 1). Just 10 minutes of meditation helps anxious people have better focus. *ScienceDaily*. Available at: www.sciencedaily.com/releases/2017/05/170501094325.htm

1 What was the aim of this investigation?

2 Write a hypothesis for this investigation.

3 Identify the independent variable.

4 Identify the dependent variable.

5 Why is a control group used in this study?

_____

_____

6 What conclusion(s) can be drawn from the results?

_____

_____

_____

7 Explain what the researchers would need to do to comply with the ethical guidelines of informed consent and voluntary participation.

_____

_____

_____

_____

_____

_____

_____

10.1.5 Community support organisations visual presentation

> **Key science skills**
> Analyse, evaluate and communicate scientific ideas
> - discuss relevant psychological information, ideas, concepts, theories and models and the connections between them
>
> *Develop*

Community support organisations such as Beyond Blue, the Black Dog Institute, Headspace and SANE Australia offer support, training and education for individuals with mental health problems and illnesses and their families. Go to these organisations' websites to find out what they do and the types of support they offer. In the space provided, create a visual display presenting a summary of the information and support programs offered by one Australian mental health and wellbeing support organisation.

10.2 Cultural determinants of social and emotional wellbeing

Key knowledge
- cultural determinants, including cultural continuity and self-determination, as integral for the maintenance of wellbeing in Aboriginal and Torres Strait Islander peoples

10.2.1 The maintenance of wellbeing in Aboriginal and Torres Strait Islander peoples

Key science skills
Analyse, evaluate and communicate scientific ideas
- discuss relevant psychological information, ideas, concepts, theories and models and the connections between them

Table 10.3 provides terms associated with the maintenance of wellbeing in Aboriginal and Torres Strait Islander peoples. Use the terms to fill in the blanks in the following paragraphs.

Each term is used only once.

Table 10.3 Terms associated with the maintenance of wellbeing in Aboriginal and Torres Strait Islander peoples

supported	responsibilities	resilience	factors
continuous	cultural determinants	protective	connecting
recognising	preservation	strengthened	sovereignty
identity	cultural continuity	practices	self-determination

1 _____ of social and emotional wellbeing are _____ factors integral to Aboriginal and Torres Strait Islander peoples' right and capacity to learn, practise and pass on their traditional and contemporary ways of knowing, being and doing. They include _____ and processes that support Aboriginal and Torres Strait Islander peoples to maintain a strong and secure sense of cultural _____ and cultural values, and to participate in cultural _____ that allow them to exercise their cultural rights and _____, and fundamental human right to self-determination.

2 _____ refers to the transmission of Aboriginal and Torres Strait Islander peoples' ways of knowing, being and doing over generations, _____ the present with the past and future. When people and communities connect with a _____ culture, their own sense of personal continuity and cultural identity is _____, which protects against the effects of intergenerational trauma and racism. Cultural continuity can be _____ through revival, renewal and _____ of both traditional and contemporary cultural knowledge and practices.

3 _____ involves recognition of the rights of Aboriginal and Torres Strait Islander peoples to practise their culture and shape their future based on their unique cultural worldview. This includes _____ Aboriginal and Torres Strait Islander _____, and issues of land rights, control of resources and cultural security. Self-determination is expressed through a proud legacy of resistance, activism and grassroots organisation and provides a source of strength and _____ for maintaining social and emotional wellbeing.

10.2.2 Aboriginal and Torres Strait Islander peoples' cultural determinants of health

Key science skills
Analyse, evaluate and communicate scientific ideas
- discuss relevant psychological information, ideas, concepts, theories and models and the connections between them

The cultural determinants of health are the protective factors that support good health and wellbeing for Aboriginal and Torres Strait Islander peoples. Complete Table 10.4 to build a summary. Provide a description of each listed cultural determinant of health and its impact on the health and wellbeing of Aboriginal and Torres Strait Islander peoples.

Table 10.4 Summary of cultural determinants of health

Cultural determinants of health	Description	Impact
Connection to *Country*		
Kinship		
Knowledge and beliefs		
Self-determination		
Cultural continuity		

10.2.3 Maintenance of mental wellbeing concepts study cards

Key science skills
Analyse, evaluate and communicate scientific ideas
- discuss relevant psychological information, ideas, concepts, theories and models and the connections between them

In this activity you will create study cards that will help you revise the key terms associated with defining mental wellbeing.

Materials

Poster cardboard, scissors, glue

What to do

1. Cut around the outside of the two columns of definitions in Table 10.5. Do not cut out each cell individually at this stage.
2. Glue the columns to your poster cardboard.
3. Now cut out the individual cardboard-backed cells of definitions. Place them in a pile and be careful not to lose them.
4. Cut out the individual cells of terms in Table 10.6. Place them in a pile and be careful not to lose them.
5. Match your cut-out definitions with your cut-out terms.
6. When they are matched, glue the paper terms to the backs of the definitions already attached to poster cardboard.
7. Use the study cards when revising this topic.

CHAPTER 10 / Maintenance of mental wellbeing

Table 10.5 Definitions related to the maintenance of mental wellbeing

Food and water intake that includes sufficient energy and nutrients to meet basic requirements for healthy living.	An emotional state characterised by the anticipation of danger, dread or uneasiness as a response to an unclear or ambiguous threat.
Authenticity is a personal trait that reflects the extent to which a person is genuine and provides social support by truly listening to and connecting with the experiences of another person.	Structured psychological techniques that can be applied to help people recognise how negative or unproductive patterns of thinking and behaviour affect their emotions; aimed at helping people think in new ways to promote more positive feelings and behaviours.
The transmission and transformation of the traditional knowledges, values and practices of a cultural group over generations, connecting the past with the present and future.	Protective factors that are integral to Aboriginal and Torres Strait Islander peoples' right and capacity to learn, practise and pass on their traditional and contemporary ways of knowing, being and doing.
The food and drink regularly provided or consumed.	In interpersonal relationships, the quality of relational energy between people refers to the capacity of a person to promote feelings of motivation, engagement and action in others.
Replenishing a lack of water in the body.	A set of techniques that are intended to encourage a heightened state of awareness and focused attention.
A state of awareness that arises when we pay non-judgemental attention to our thoughts and feelings in the present moment.	A meditation practice in which a person focuses attention on their breathing, with thoughts, feelings and sensations being experienced freely as they arise and without judgement.
The daily eating patterns of an individual, including specific foods and calories consumed and relative quantities.	Any behavioural, biological, psychological or environmental characteristic that decreases the likelihood of a person developing a particular mental health problem or disorder.
A person's ability to successfully adapt to stress and cope with adversity, influenced by coping strategies, adaptive ways of thinking and social connectedness.	The genetic and environmental conditions that influence the likelihood that a person will experience a mental health condition or another negative health outcome.
The assistance and comfort we receive from people in our social network when we are facing a stressful or challenging situation, from family members and friends through to support groups and social institutions.	The fundamental right of Aboriginal and Torres Strait Islander peoples to shape their own lives, so that they determine what it means to live well according to their own values and beliefs.

Table 10.6 Terms related to the maintenance of mental wellbeing

Adequate diet	Anxiety
Authentic social support	Cognitive behavioural strategies
Cultural continuity	Cultural determinants
Diet	Energising social support
Hydration	Meditation
Mindfulness	Mindfulness meditation
Nutritional intake	Protective factor
Resilience	Risk factor
Self-determination	Social support

Table 10.6 Terms related to the maintenance of mental wellbeing

Exam practice

Multiple choice

Circle the response that best answers the question.

1 A person's nutritional intake will not _____ the development of a mental illness; however, it can help to _____ good mental wellbeing and contribute to a person's level of resilience when presented with adversity.
   A prevent; cause
   B promote; prevent
   C prevent; promote
   D cause; prevent

2 Eating a balanced diet can help
   A improve mood.
   B diminish brain function.
   C reduce resilience.
   D all of the above.

3 The Australian Dietary Guidelines recommends that women should have about _____ cups and men about _____ cups of water per day to stay adequately hydrated.
   A 6; 8
   B 7; 9
   C 8; 10
   D 10; 12

4 The amount of adequate sleep will differ between individuals depending on which of the following factors?
   A age
   B physical activity levels
   C general health
   D all of the above

5 Researchers believe the benefits of mindfulness are related to
   A reduction in the release of the neurotransmitter dopamine.
   B its ability to dial down the body's response to stress.
   C the activation of the flight-or-fight-or-freeze response.
   D hyperactivity in the amygdala.

6 The revival of Indigenous Australian languages, art, songs and stories aid the restoration of
   A cultural continuity.
   B self-determination.
   C political continuity.
   D social continuity.

7 The fundamental right of people to shape their own lives, so that they determine what it means to live well according to their own values and beliefs, is known as
   A freedom.
   B cultural responsibility.
   C historical authority.
   D self-determination.

8 Aboriginal and Torres Strait Islander peoples experience _____ times more disease-related illness than non-Indigenous Australians.
- A 1.5
- B 2.0
- C 2.3
- D 3.0

Short answer

1 Explain how a person's nutritional intake may influence their mental wellbeing 3 marks

2 What is the aim of cognitive behavioural strategies? 2 marks

3 What is cultural continuity and how may it impact the wellbeing of Aboriginal and Torres Strait Islander peoples? 3 marks

Using scientific inquiry   11

11.1 Designing an investigation

Key knowledge
Investigation design
- psychological concepts specific to the selected scientific investigation and their significance, including definitions of key terms
- characteristics of the selected scientific methodology and method, and appropriateness of the use of independent, dependent and controlled variables in the selected scientific investigation
- techniques of primary quantitative data generation relevant to the selected scientific investigation
- the accuracy, precision, repeatability, reproducibility and validity of measurements
- the health, safety and ethical guidelines relevant to the selected scientific investigation

11.1.1 Formulating hypotheses

Key science skills
Develop aims and questions, formulate hypotheses and make predictions
- identify independent, dependent and controlled variables in controlled experiments
- formulate hypotheses to focus investigations

Practise

This activity will give you practice in predicting the outcome of an experiment. To do this, you will formulate a testable hypothesis.

Read the scenarios below. For each scenario you will be asked to:
- identify the independent variable
- identify the dependent variable and how it will be measured
- formulate a research hypothesis that includes your IV and DV.

Remember, when you are formulating a hypothesis it has to include the IV and how it will be manipulated, the DV and how it will be measured, and a prediction of the relationship between the two.

SCENARIO 1

Dr Sing wants to research the influence of nutrient-dense 'super foods' such as blueberries and purple grapes on age-related memory decline. For two months the experimental group is given a measured dose of super foods and then they are tested on the number of words they recall on a series of short word-recollection tests. The control group completes the same test but does not consume any super foods in the months prior to doing the test.

1 Identify the independent variable.

2 Identify the dependent variable and how it will be measured.

3 Formulate a research hypothesis that includes your independent and dependent variables.

SCENARIO 2

A researcher is interested in the effect of pregnant mothers drinking alcohol on the birth weight of babies. She selects a sample of 50 pregnant rats and adds alcohol to the food of 25 of the rats once a week, to simulate weekend drinking. One pup is randomly selected from each litter and is weighed 24 hours after birth.

1 Identify the independent variable.

2 Identify the dependent variable and how it will be measured.

3 Formulate a research hypothesis that includes your independent and dependent variables.

SCENARIO 3

It is believed that texting while driving increases the number of accidents, particularly in P-plate drivers. To test this, a researcher set up an obstacle course in a supermarket carpark. A number of P-plate drivers were asked to complete the obstacle course as quickly as possible, while receiving messages and texting a response. A week later, the same drivers were asked to complete the course without texting. The researcher recorded the number of cones hit by the drivers.

1 Identify the independent variable.

2 Identify the dependent variable and how it will be measured.

3 Formulate a research hypothesis that includes your independent and dependent variables.

SCENARIO 4

Dr Asch, a psychologist, was concerned about the lack of self-esteem in adolescents. She wondered if adolescents who belonged to a sporting club had higher self-esteem. Dr Asch gave a questionnaire to a sample of 200 adolescents who had been members of a sporting group for at least three years. She gave the same questionnaire to 200 adolescents who had not been a member of a sporting group in the last three years. She created a score for each answer and then compared the scores from the two groups.

1 Identify the independent variable.

2 Identify the dependent variable and how it will be measured.

3 Formulate a research hypothesis that includes your independent and dependent variables.

SCENARIO 5

For his PhD, a university student investigated the effects of heat on cognitive performance. The student divided the participants into three groups: group one was placed into a room at 25 °C, group two was in a room at 30 °C and group three was in a room at 35 °C. Each group was given a general knowledge test to complete in 20 minutes.

1   Identify the independent variable.

2   Identify the dependent variable and how it will be measured.

3   Formulate a research hypothesis that includes your independent and dependent variables.

SCENARIO 6

A group of researchers wanted to investigate the effect of a drug on sleep quality in patients suffering from PTSD. Patients were asked to record the number of times they woke up during the night for one month. They were then given the new drug for two months. After this period, they were asked to again record the number of times they woke up during the next month.

1   Identify the independent variable.

2   Identify the dependent variable and how it will be measured.

3   Formulate a research hypothesis that includes your independent and dependent variables.

11.1.2 Extraneous variables of all shapes and sizes

Key science skills
Analyse and evaluate data and investigation methods
- identify and analyse experimental data qualitatively, applying where appropriate concepts of: accuracy, precision, repeatability, reproducibility and validity; errors; and certainty in data, including effects of sample size on the quality of data obtained
- evaluate investigation methods and possible sources of error or uncertainty, and suggest improvements to increase validity and to reduce uncertainty

Develop

This activity will give you practice in identifying extraneous variables that may occur in research. It will also help you decide how to eliminate them from further research.

Read the scenarios below. For each scenario you will be asked to:
- identify two potential extraneous variables that may affect the research scenario
- suggest how the extraneous variables could be controlled.

Make sure you understand exactly what an extraneous variable is before you start this activity.

SCENARIO 1

Alex, a developmental psychologist, was investigating whether preschool-aged children prefer a particular colour. She showed 100 children a palette with the colours red, green and blue on it and then timed how long each child spent looking at each colour.

1. One extraneous variable:

2. How to control this extraneous variable:

3. Another extraneous variable:

4. How to control this extraneous variable:

SCENARIO 2

A teacher wonders whether there is a link between eating meat and IQ. She separates a sample of high school students into those who eat meat and those who do not eat meat. Each group completes an IQ test.

1. One extraneous variable:

2. How to control this extraneous variable:

3. Another extraneous variable:

4. How to control this extraneous variable:

SCENARIO 3

A music teacher wanted to find out whether students who learn a musical instrument are more successful in Year 12. She collected the ATAR scores for the last three years and divided them into students who had studied music and those who had not.

1. One extraneous variable:

2. How to control this extraneous variable:

3. Another extraneous variable:

4. How to control this extraneous variable:

SCENARIO 4

John is a student teacher who has been given an assignment to research whether older teachers can motivate students as well as younger teachers can. When out on teaching rounds, he observes a number of teachers in class, then gives the students a self-report to complete.

1. One extraneous variable:

2. How to control this extraneous variable:

3. Another extraneous variable:

4 How to control this extraneous variable:

_____

_____

11.1.3 Which experimental design is which?

Key science skills
Plan and conduct investigations
- determine appropriate investigation methodology: case study; classification and identification; controlled experiment (within subjects, between subjects, mixed design); correlational study; fieldwork; literature review; modelling; product, process or system development; simulation

Analyse and evaluate data and investigation methods
- evaluate investigation methods and possible sources of error or uncertainty, and suggest improvements to increase validity and to reduce uncertainty

Analyse, evaluate and communicate scientific ideas
- discuss relevant psychological information, ideas, concepts, theories and models and the connections between them

This activity will allow you to identify and evaluate different experimental designs.

PART A

Complete Table 11.1 with descriptions for the experimental designs, as well as the advantages and disadvantages of each.

Table 11.1 Evaluating experimental designs

Experimental design	Description	Advantages	Disadvantages
Independent-groups			
Matched-participants			
Repeated-measures			

9780170465069

PART B

Draw lines in Table 11.2 to join each scenario to the most appropriate experimental design.

Table 11.2 Evaluating appropriateness of experimental designs

Scenario	Experimental design
A study with a small number of participants	
A study that compares before and after scores	Independent-groups design
A study where experimenters have little money	
A study using identical twins	
A study that seeks to compare two groups, one of which is exposed to the experimental condition	Matched-participants design
An important study of a technique to improve learning	
A study requiring a comparison	Repeated-measures design

11.1.4 Sampling procedures

Key science skills

Plan and conduct investigations
- design and conduct investigations; select and use methods appropriate to the investigation, including consideration of sampling technique (random and stratified) and size to achieve representativeness, and consideration of equipment and procedures, taking into account potential sources of error and uncertainty; determine the type and amount of qualitative and/or quantitative data to be generated or collated

Analyse and evaluate data and investigation methods
- evaluate investigation methods and possible sources of error or uncertainty, and suggest improvements to increase validity and to reduce uncertainty

Analyse, evaluate and communicate scientific ideas
- discuss relevant psychological information, ideas, concepts, theories and models and the connections between them

Develop

This activity will allow you to investigate different types of sampling techniques and summarise some of the steps involved in conducting psychological research.

PART A

Use terms from Table 11.3 to fill the gaps in the paragraphs that follow.

Each term is used only once.

Table 11.3 Data terms

large	generalised	sample	strata	biased
quick	time	random	representative	subgroup
results	opportunity	numbers	stratified	selection

1. The process of choosing participants for a study is called participant _____. It is often not possible to include all members of a population of interest, so a _____ must be taken from the population. A population is a group of people to which the research is relevant. For example, the population could be all people who suffer from depression. The sample is a _____ of the population and needs to be _____ of that population. If it isn't, the results cannot be _____ from the sample to the population. To make the sample representative, the experimenter must ensure that all members of the population have an equal chance of being selected. This is done using _____ sampling. This method could involve putting the names of all members of the population into a container and then pulling out a set number of names. Each name that is pulled out becomes part of the sample.

A table of random _____ may also be used to create the sample. This sampling method improves the chances of making accurate inferences based on the results, but it may not always be possible to obtain a list of all members of a population, particularly if the population is _____.

2 _____ sampling allows a researcher to create a sample in the same proportions that are found in the population. The population is divided into groups, called _____, which share something in common. For example, a population may be divided into age groups. Each age group can then be randomly sampled. While this procedure makes the sample more representative, it is very _____ consuming.

3 Convenience sampling is also known as _____ sampling. Here, a researcher chooses participants who are readily available, so this is a _____, inexpensive way for a researcher to select a sample. For example, a researcher based at a university uses university students to make up their sample. Other examples of convenience sampling include newspaper, radio and phone polls. The risk with this type of sampling is that the sample may be _____ and not representative of the wider population, so any _____ may not be able to be generalised.

PART B

Provide definitions for each of the stages shown in the flowchart in Figure 11.1. Add advantages and disadvantages where required. It may help to increase the size of this flowchart by photocopying it onto an A3 sheet of paper.

Figure 11.1 Psychological research flowchart

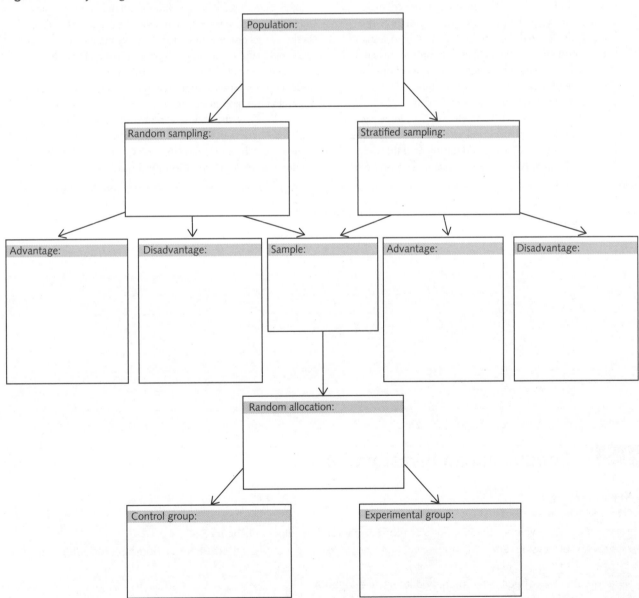

11.1.5 Ethics in research

Key science skills

Analyse, evaluate and communicate scientific ideas
- discuss relevant psychological information, ideas, concepts, theories and models and the connections between them

Comply with safety and ethical guidelines
- demonstrate ethical conduct and apply ethical guidelines when undertaking and reporting investigations

Develop

This activity will reinforce your understanding of different ethical guidelines and concepts in psychological research, as well as help you to identify correct and incorrect ethical procedures.

1. Modern-day experiments are governed by a code of ethics, but this was not so in the past. Read the following article about an experiment that today would be considered unethical.
2. Complete the task that follows the article.

> In 1939, Wendell Johnson was an assistant professor at the University of Iowa, known for its research on stuttering. He supervised a graduate student, Mary Tudor, who conducted the research. He believed that he could cause children to stutter, or to stop stuttering, depending on how the children were treated. He would then be able to conclude that stuttering is not inherited, as was commonly believed. This conclusion would have had huge implications for speech therapy, a new discipline at the time. The research later became known as the Monster Study.
>
> First, Johnson and Tudor selected 22 children, aged 5 to 15, from a local orphanage and split them into experimental and control groups. None was told the intent of the research. Of the 22 children, 10 were chosen because they had been classified as stutterers by matrons at the orphanage. Half of this group were allocated to the experimental group and half to the control group. The remaining 12 children were randomly chosen. Tudor used kindness with the control group, praising the fluency of their speech, even if they stuttered. The children in the control group who started the experiment with a stutter were told that the stuttering was only a phase and that they would grow out of it. With the experimental group, Tudor picked up every speech imperfection and constantly reminded the children to stop stuttering.
>
> Many of the children in the experimental group suffered ongoing psychological effects, some never losing their stutter. In 2001, the University of Iowa publicly apologised for Johnson's Monster Study.

1. Name the ethical concepts and guidelines that have been breached.

_____
_____
_____
_____
_____
_____
_____
_____
_____
_____

11.2 Conducting an investigation

Key knowledge

Scientific evidence
- the nature of evidence that supports or refutes a hypothesis, model or theory
- ways of organising, analysing and evaluating primary data to identify patterns and relationships, including sources of error and uncertainty
- authentication of generated primary data using a logbook
- assumptions and limitations of investigation methodology and/or data generation and/or analysis methods
- criteria used to evaluate the validity of measurements and psychological research

11.2.1 Experimental data

Key science skills

Plan and conduct investigations
- design and conduct investigations; select and use methods appropriate to the investigation, including consideration of sampling technique (random and stratified) and size to achieve representativeness, and consideration of equipment and procedures, taking into account potential sources of error and uncertainty; determine the type and amount of qualitative and/or quantitative data to be generated or collated

Generate, collate and record data
- systematically generate and record primary data, and collate secondary data, appropriate to the investigation
- record and summarise both qualitative and quantitative data, including use of a logbook as an authentication of generated or collated data
- organise and present data in useful and meaningful ways, including tables, bar charts and line graphs

Analyse, evaluate data and investigation methods
- evaluate investigation methods and possible sources of error or uncertainty, and suggest improvements to increase validity and to reduce uncertainty

In this activity you will demonstrate your understanding of types of data and ways of obtaining data in psychological research.

PART A

Read the following scenarios and answer the questions that follow each.

SCENARIO 1

A researcher wearing a white coat and carrying a clipboard stands outside a primary school at the end of a school day. She records the number of drivers who slow to the required 40 km/h as they pass the school.

1 For how many days would the researcher need to be present to get a representative sample of driver behaviour?

2 What would be a control condition for this experiment?

3 Identify the type of study being conducted.

4 Is the data collected qualitative or quantitative?

SCENARIO 2

Approximately 9 months after having a stroke, Harry noticed that if he touched something it would stimulate his tastebuds. Words written in a certain shade of green made him feel happy. Yellow created feelings of disgust. He was diagnosed by a psychologist as having synaesthesia. This is a condition that causes a person's senses, which are usually experienced alone, to join with other senses to produce unusual effects. Harry was studied by the psychologist for the next 5 years.

1 Name the data collection technique that relates to Harry.

2 Would the data collected by the psychologist be qualitative, quantitative or both? Explain your answer.

3 This study of Harry was conducted over several years. What is the name for such a study?

4 The data on Harry was added to data collected about other people with synaesthesia. What is one reason for doing this?

SCENARIO 3

Gillian, a psychologist, wanted to know what proportion of nurses employed in Victoria were male.

1   Suggest how Gillian could obtain a representative sample for this study.

2   What data collection technique is likely to be used?

3   Suggest how Gillian could present the data that she collects.

PART B

This activity provides you practice in summarising and organising your data.

1   Infant birth weights (in kilograms) at a country hospital for the months of January and July were as follows:

   January: 2.8, 3.7, 3.9, 3.6, 2.9, 2.5, 3.5, 3.2, 3.0, 3.8, 2.7

   July: 2.5, 2.4, 2.8, 3.2, 3.5, 4.4, 3.9, 3.8, 2.6, 2.6, 3.0

   a   Calculate the mean (average) birth weight for each month.

   b   Create a frequency table for this data.

c Compare the spread of birth weights in January and July, using your frequency table.

_____

_____

_____

2 What kind of graph would you use to show the growth rate of two groups of rats, one of which has been exposed to soothing music, while the other has not? Explain your choice.

_____

_____

_____

_____

Make sure you follow the conventions for drawing a scientific graph.

3 The students in Mr Payne's class voted on their favourite colour. Each student voted anonymously once. The results were: 8 votes for purple, 7 votes for blue, 5 votes for red, 2 votes for green and 3 votes for no favourite. Mr Payne converted these figures to percentages. Create an appropriate graph (using the grid paper below) using the percentages that Mr Payne would have calculated.

11.2.2 Applying your knowledge

> **Key science skills**
> Develop aims and questions, formulate hypotheses and make predictions
> - identify independent, dependent and controlled variables in controlled experiments
> - formulate hypotheses to focus investigations
>
> Analyse and evaluate data and investigation methods
> - identify and analyse experimental data qualitatively, applying where appropriate concepts of: accuracy, precision, repeatability, reproducibility and validity; errors; and certainty in data, including effects of sample size on the quality of data obtained
> - evaluate investigation methods and possible sources of error or uncertainty, and suggest improvements to increase validity and to reduce uncertainty
>
> Construct evidence-based arguments and draw conclusions
> - evaluate data to determine the degree to which the evidence supports the aim of the investigation, and make recommendations, as appropriate, for modifying or extending the investigation
> - evaluate data to determine the degree to which the evidence supports or refutes the initial prediction or hypothesis
> - use reasoning to construct scientific arguments, and to draw and justify conclusions consistent with evidence base and relevant to the question under investigation
> - identify, describe and explain the limitations of conclusions, including identification of further evidence required
>
> Analyse, evaluate and communicate scientific ideas
> - discuss relevant psychological information, ideas, concepts, theories and models and the connections between them

Develop

In this activity you will apply your knowledge and understanding of research skills to a scenario.

Read the scenario below and then answer the questions that follow.

SCENARIO

A psychologist wanted to test whether or not stress levels in VCE students changed during the exam period. She advertised in a local newspaper for participants currently enrolled in VCE and sampled 100 of the applicants by drawing their names out of a hat.

Students were asked to rate their stress on a 100-point scale in the week before, the week of and the week after their end-of-year exams.

The results showed that, during the week before exams, the participants' average stress rating was 84. During the exam week, the average rating was 89 and in the week after the exams the average rating was 20.

1 What methodology was used in this investigation? Explain your answer.

2 What is the independent variable for this investigation?

3 What is the dependent variable for this investigation?

4 a What is a research hypothesis?

b Write a research hypothesis for this investigation.

5 a What type of sampling procedure was used in this investigation?

b How does this type of sampling procedure affect the investigation?

6 a What type of experimental design is used in this investigation?

b One disadvantage of this experimental design is the presence of order effects. What is an order effect? Provide an example.

c What is one way that order effects can be controlled? Explain this technique.

7 Can the experimenter generalise from this investigation to other stressful situations? Explain your answer.

11.3 Science communication

Key knowledge
Science communication
- conventions of science communication: scientific terminology and representations, symbols, formulas, standard abbreviations and units of measurement
- conventions of scientific poster presentation, including succinct communication of the selected scientific investigation and acknowledgements and references
- the key findings and implications of the selected scientific investigation

11.3.1 Analysing research

> **Key science skills**
> Analyse and evaluate data and investigation methods
> - evaluate investigation methods and possible sources of error or uncertainty, and suggest improvements to increase validity and to reduce uncertainty
>
> Analyse, evaluate and communicate scientific ideas
> - discuss relevant psychological information, ideas, concepts, theories and models and the connections between them

This activity will help you to identify common errors that are made when writing a scientific report.

PART A

1. Read the example of a report on an investigation and highlight or make notes about errors that have been made in writing style or formatting.
2. Create a list of the mistakes you have identified.

The effect of reinforcement on submission of homework

ABSTRACT

The aim of this study was to test the effect of reinforcement on the amount of homework submitted on time. I hypothesised that Group 1 would submit more homework than Group 2. The participants used in the study were 40 male and female students; 20 in each group. The results showed that Group 1, which was praised for handing in homework, handed in 92 per cent of homework tasks and Group 2, which received no reinforcement for handing in homework, handed in 73 per cent of homework tasks. In conclusion, it can be seen that Group 1 handed in more homework than Group 2.

INTRODUCTION

Operant conditioning is a form of learning that relies upon the consequences of a behaviour to guide future actions (2003). Behaviour that has satisfying consequences, such as praise, will lead to an increase in that behaviour occurring again, and behaviour that has negative consequences, such as being grounded, will lead to a decrease in the likelihood of that behaviour occurring again.

One type of consequence is known as positive reinforcement. Positive reinforcement is the giving of a pleasant stimulus to increase the likelihood of a behaviour occurring again (2010).

The aim of this study was to test the effect of reinforcement on the amount of homework submitted.

I hypothesised that Group 1, which was praised for handing in homework, would submit more homework tasks than Group 2, which received no reinforcement for handing in homework.

The independent variable in this study was the percentage of homework tasks submitted. The dependent variable was the presence of positive reinforcement.

Method

PARTICIPANTS

The participants used in the study were 40 male and female students; 20 in each group.

MATERIALS

The following materials were used to conduct the experiment:
- three homework tasks (40 copies of each)
- 40 participants

PROCEDURE

Participants were selected from one school and broken into two sets of 20 students using random allocation. Participants were asked to complete homework assignment 1 and submit it the following day. The experimenter would praise all students in Group 1 who submitted the work. The experimenter would not praise any students in Group 2. This was then repeated over the next two days with homework assignments 2 and 3.

RESULTS

Figure 11.2 Results graph

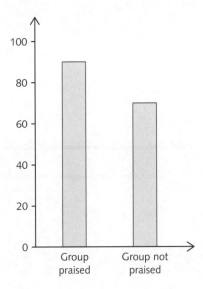

The graph above shows the percentage of homework tasks submitted by Group 1 and Group 2.

DISCUSSION

I hypothesised that Group 1 would submit more homework tasks than Group 2.

The hypothesis was supported, as the group that was praised submitted a higher percentage of homework tasks than the group that was not praised. These findings support past research.

A possible extraneous variable that could have affected these results was that one class may have been more studious than the other. In the future a repeated-measures design should be used so that the same students are exposed to both the control and experimental conditions.

Another possible extraneous variable was how the students felt about the teacher. In the future a survey on the students' feelings towards the teacher should be conducted and then the sample could be matched based on their feelings towards the teacher.

In conclusion, it can be seen that praise led to an increase in homework tasks submitted. This study demonstrates that positive reinforcement may lead to an increase in desired behaviour.

REFERENCES

K. Robinson (2003). Psychology basics.

1 List all the mistakes you found in this report.

PART B

You are going to use the report above to create a scientific poster for assessment.
1 Write a communication statement for the investigation above. Make sure the statement is 15 words or fewer.

2 Redraw the graph in Figure 11.2 so it follows the conventions for a scientific graph.

3 Write a conclusion for the investigation above. Make sure the conclusion is 40 words or fewer.

_____
_____
_____

PART C

Place a tick in the correct column of Table 11.4 to indicate whether each statement is true or false.

Table 11.4

	True	False
A scientific report should be written in past tense and a poster in the present tense.		
The results section of a poster can include tables and figures.		
The title of the book appears first when referencing using APA format.		
The aim and the hypothesis appear in the introduction on a poster.		
The results section of a poster can include a bullet list of results.		
A poster that presents research should be scientific in style, not personal.		
A conclusion is required on a scientific poster.		
The participants should be mentioned in the materials section.		
When referencing quotes in the body of the text, only the author's surname and year of publication are required.		
When drawing a graph, no title is necessary because the section is already titled 'Results'.		

11.3.2 Scientific terminology

> **Key science skills**
> Analyse, evaluate and communicate scientific ideas
> - discuss relevant psychological information, ideas, concepts, theories and models and the connections between them

Materials
Poster cardboard, scissors, glue

Instructions
1. Study the definitions in Table 11.5 and the terms in Table 11.6.
2. Cut around the outside of the two columns of definitions in Table 11.5. Do not cut out each cell individually at this stage.
3. Glue the columns to your poster cardboard.
4. Cut out the individual cardboard-backed rectangles of definitions. Be careful not to lose any.
5. Cut out the individual rectangles for the terms in Table 11.6.
6. Match each term to its correct cardboard-backed definition and glue them back-to-back.

Table 11.5 Definitions related to key science skills

In science, the accuracy of a measurement relates to how close it is to the 'true' value of the quantity being measured. Accuracy is not quantifiable; measurement values may be described as more accurate or less accurate.	In research, the aim is the purpose of the study (what you intend to investigate).
A chart that visualises data using separated rectangular columns or 'bars' to represent the total number, mean or other measures for distinct categories of data.	A research investigation design in which groups of participants or sets of data are unrelated to one another, often comparing a control group to an experimental group; also called independent-groups design.
A research investigation that focuses on a particular person, activity, behaviour, event or problem that is, or could be, experienced within a real-world context outside of the laboratory; may involve direct observation or may be a historical analysis of causes, consequences and implications for the development of knowledge.	The tendency for a majority of scores to fall in the mid-range of possible values in a data set; measured by a range of descriptive statistics such as the mean, mode and median.
The component of a research report that describes the outcomes or findings of the investigation and the implications of the results for the research aims.	In research ethics, the requirement to ensure the privacy, protection and security of a participant's personal information and the anonymity of individual results.

CHAPTER 11 / Using scientific inquiry

In an experiment, an unintended variable, distinct from the intended independent variable(s), that affects the dependent variable in the same way as the hypothesised effect of the independent variable(s), making it impossible to determine the cause of the results; that is, the results are confounded.	The comparison group in a scientific investigation who do not experience the treatment, intervention or experimental condition; ideally, matched to the experimental group on other relevant variables that may affect the results.
A kind of scientific investigation in which a researcher systematically manipulates one or more independent variables to determine the effect on an outcome variable(s) (dependent variable) while attempting to control (eliminate or neutralise) the influence of other variables that may also affect the dependent variable, with the goal to determine cause and effect (causation).	A variable that is not of central interest to a scientific investigation (i.e. extraneous) that the researcher seeks to control using a particular strategy (e.g. matching groups) so as to eliminate or neutralise its potential effect on the results.
A method for selecting participants for a study that is based on ease of access for the researcher.	A measure of the strength of relationship between two variables, ranging from −1 to +1; a zero correlation indicates that there is no relationship; −1 indicates a perfect negative correlation; +1 indicates a perfect positive correlation. It is important to remember that even a perfect correlation does not indicate a causal relationship between the two variables.
A scientific investigation that involves measuring two or more variables that have not been manipulated or controlled to understand the relationships/associations that exist between them (e.g. height and weight); used to identify which variables may be more important to study further and to make predictions.	A method used by researchers to control order effects by systematically exposing different participants to different orders of treatment conditions.
The observations or measurements that are obtained and recorded when undertaking a research investigation.	The process at the end of a research study in which participants are given a complete explanation of the study, which may not have been possible before or during the research, and an opportunity to seek clarification; essential in studies that use deception.
When participants are deliberately misled or not fully informed of the true nature or purpose of a research investigation; appropriate when knowledge of the true nature of the study would likely influence participants' responses and threaten internal validity.	The outcome measure of interest in a scientific investigation that is predicted to be affected by the independent variable(s); represented in graphs on the vertical (Y) axis.
Numerical values and methods for visualising these that summarise the main features of a data set, including measures of central tendency (mean, mode, median) and of the range and spread of the data (standard deviation); visualised in charts, tables and graphs.	An experimental procedure in which neither the experimenter nor the participants know to which experimental condition the participants have been allocated.
Judgements and decisions guided by principles and values about how we should act to promote the most beneficial outcomes.	The branch of philosophy that asks questions and provides guidance about the principles and values we should apply in making judgements and decisions about how to act.

9780170465069

CHAPTER 11 / Using scientific inquiry

In a controlled experiment, the group of participants that are exposed to the independent variable.	A statement of a testable prediction about the effect of one variable on another or about the relationship between variables, usually based on theory.
Any expectations, beliefs or preferences of a researcher that may unintentionally influence their study design choice, recording of observations and/or their interactions with participants in such a way as to affect the outcome of the investigation.	The extent to which the results of a scientific investigation can be applied (generalised) to other people or situations beyond the sample and context used in the investigation.
Any variable other than the independent variable that may affect the results (dependent variable) of the research.	A form of research investigation in which the researcher observes and interacts with a selected environment beyond the laboratory, usually to determine correlation rather than a causal relationship.
The extent to which research findings can be applied to people and/or contexts other than those involved in the research; see also external validity.	A testable prediction about the relationship between two variables.
The factor or condition (variable) that an experimenter manipulates (changes or varies) systematically to determine its effect on another variable (the dependent variable).	A fundamental component of ethical research in which, at the beginning of the research process, the researcher seeks a person's voluntary agreement to participate in a scientific investigation based on having ensured that the person fully understands the purpose and nature of the methods and procedures, the potential risks and benefits, and their right to withdraw at any stage without penalty.
In ethical research, the commitment to searching for knowledge and understanding, and the honest reporting of all sources of information and results, whether favourable or unfavourable, in ways that permit scrutiny and contribute to public knowledge and understanding.	A measurement scale in which the distance between any two values is equal with no natural zero point.
A form of qualitative data collection where people are asked to comment on their attitude towards particular issues.	A study that collects data over two or more periods in time, using the same participants.
An experimental design in which participants are paired (matched) on the basis of similar characteristics that can influence the dependent variable, with one of the pair being allocated to the experimental group and the other to the control group.	A measure of central tendency that gives the numerical average of a set of scores, calculated by adding all the scores in a data set and then dividing the total by the number of scores in the set.

9780170465069

A type of descriptive statistic that summarises the typical or average value of a set of scores, usually in the form of the mean (average), median or mode.	A measure of central tendency, calculated by arranging scores in a data set from the highest to the lowest and selecting the middle score.
A research design that includes both within- and between-subjects conditions as independent variables.	A measure of central tendency found by selecting the most frequently occurring score in a set of scores.
In ethical research, to avoid causing harm. However, as positions or courses of action in scientific research may involve some degree of harm, this concept implies that the harm resulting from any position or course of action should not be disproportionate to the benefits from any position or course of action.	An informal measurement instrument used to assess cognitive performance or skills that does not have strictly controlled procedures for administration or any formally determined standards (norms) against which to compare a person's performance.
A kind of scientific investigation that involves watching and recording the behaviour of other persons or animals within a specific environment without intervening or manipulating variables.	A form of bias that occurs when a researcher's expectations, past experiences, motives or other personal factors interfere with the accuracy of their observations; see also experimenter effect.
When a person's behaviour changes because they are aware that they are being observed; also called the Hawthorne effect.	A precisely defined and described prediction about what will happen when an independent variable is manipulated and the effect this is expected to have on the dependent variable.
Data points within a data set that are distant from the majority of other values in the data set; their impact on the data needs to be carefully considered when describing the outcomes of a scientific investigation.	Individual differences in the personal characteristics of research participants that can confound the results of the experiment if not controlled.
The people or animals whose behaviour, characteristics or responses are investigated and measured as part of a scientific investigation.	The personal rights of all participants that must be respected by the researcher, as outlined in ethical guidelines relating to psychological research.

When someone experiences a clinically significant change in symptoms after receiving a treatment or substance known to have no such effect that occurs because of expectations associated with receiving treatment.	In human research, the entire group of people that is of interest to a researcher from which they draw a sample.
A link between two variables that indicates that increases in the value of one variable are related to increases in the level of the other.	Refers to how close a set of measurement values are to one another; determined by the repeatability and/or the reproducibility of the measurements obtained using a particular measurement instrument and procedure.
Information collected for a research project that records or describes the attitudes, behaviours or experiences of participants conceptually rather than through numerical measurement techniques (e.g. data collected through questionnaires or interviews).	Information collected in a scientific investigation in the form of carefully controlled numerical measurements.
A set of questions designed to obtain information about a person's attitudes, beliefs, feelings, behaviours, abilities, experiences or other characteristics, which may require written responses, interviews or interaction with a website.	A procedure for assigning participants to either the experimental group or control group in an experiment, ensuring that all participants have an equal chance of being allocated to either group.
A sampling technique that uses a chance process to ensure that every member of the population of interest has an equal chance of being selected for the sample.	A sampling technique in which sub-groups (strata) are identified within the population of interest and a chance procedure is used to select participants from within each sub-group, ensuring equal chance of selection for people within each sub-group and matching the number of participants from each sub-group to the proportion of the sub-group within the population.
The difference between the highest score and the lowest score in a distribution of scores.	A research tool that assigns a number on a scale to indicate the degree of agreement with a statement or the frequency of some behaviour.
A measurement scale in which data are categorised into groups and ranked with even intervals; has an absolute zero.	The information collected in a research investigation in its initial form, before being screened for outliers or invalid responses or otherwise processed.

When writing an essay or report, to acknowledge a concept, definition or idea taken from another source.	The extent to which a questionnaire, scale or any other psychological measure provides consistent results across multiple applications.
The closeness of the agreement between the results of successive measurements of the same quantity being measured, carried out under the same conditions of measurement.	A randomly selected group of participants that accurately reflects the characteristics of the larger population from which it is drawn.
The closeness of the agreement between the results of measurements of the same quantity being measured, carried out under changed conditions of measurement.	When the results can be generalised to relevant populations beyond the sample.
A formal scientific study or experiment that is designed to answer a specific research question, reported using standard writing and analysis conventions in a research report or scientific paper.	In human research, the group of people selected to participate in a research investigation from a population of interest.
The process of selecting participants from a population of interest to provide the data for a research investigation.	An approach to generating knowledge that is based on a set of organised and systematic procedures, principles, attitudes and behaviours for collecting, interpreting and verifying data, with the aim to produce reproducible evidence that forms the basis for scientific laws, principles, theories and models that generate hypotheses that can be tested.
A method for collecting data about people's behaviours, attitudes or traits that cannot be easily directly observed that relies on participants answering questions honestly that relate to their feelings, beliefs, attitudes or behaviours; can be affected by self-serving bias and socially desirable responses.	The use of clearly specified procedures for administering and scoring tests to ensure that results from different tests can be compared and interpreted meaningfully.
A test that has been developed with clearly specified procedures for administration and scoring to ensure that the results from different attempts on the test can be compared and interpreted meaningfully.	A sampling technique used to ensure that a sample contains the same proportions of participants from each social level or group (strata) that exists in the population of interest.
An extraneous variable whose influence has not been eliminated from an experiment because the experimenter was not aware of it.	The extent to which the design of a scientific investigation and the measurements chosen (e.g. a questionnaire, errors or response times) provide meaningful and generalisable information about the psychological constructs of interest.

A summary statistic (numerical value) that tells us the degree to which scores in a distribution are spread out or clustered together, such as the standard deviation from the mean.	Any condition (stimulus, event, quality, trait or characteristic) that can take a range of values that can be measured or manipulated in a scientific investigation.
When participants willingly agree to take part in a research investigation free from pressure or fear of negative consequences and understanding what they will be asked to do.	The entitlement (right) of a person who has consented to participate in a research investigation to be able to leave the study (withdraw) at any stage, without explanation and without negative consequences.
When the study design and methods include effective measures of the psychological constructs of interest so that the study's results can be interpreted meaningfully in relation to the aims of the study.	

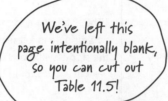

Table 11.6 Terms related to key science skills

Accuracy	Aim
Bar graph	Between-subjects design
Case study	Central tendency
Conclusion	Confidentiality
Confounding variable	Control group
Controlled experiment	Controlled variable
Convenience sampling	Correlation
Correlational study	Counterbalancing
Data	Debriefing
Deception	Dependent variable (DV)

Descriptive statistics	Double-blind procedure
Ethical considerations	Ethics
Experimental group	Experimental hypothesis
Experimenter effect	External validity
Extraneous variable	Fieldwork
Generalisability	Hypothesis
Independent variable (IV)	Informed consent
Integrity	Internal validity
Interval scale	Interview
Longitudinal study	Matched-participants design

Mean	Measure of central tendency
Median	Mixed design
Mode	Non-maleficence
Non-standardised test	Observational studies
Observer bias	Observer effect
Outlier	Participant variables
Participants	Participants' rights
Placebo effect	Population
Positive correlation	Precision
Qualitative data	Quantitative data

Questionnaire	Random allocation
Random sampling	Random stratified sampling
Range	Rating scale
Ratio scale	Raw data
Reference	Reliability
Repeatability	Representative sample
Reproducibility	Research hypothesis
Sample	Sampling
Scientific investigation	Scientific method
Self-report	Standardisation

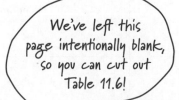

Standardised test	Stratified sampling
Uncontrolled variable	Validity
Variability	Variable
Voluntary participation	Withdrawal rights

Exam practice

Multiple choice

Circle the response that best answers the question.

1. Dr Anderson wanted to investigate the effects of caffeine on concentration. Which one of the following would be the best population for this study?
   A. all students under the age of 25
   B. all adults between the ages of 18 and 50 years who consume caffeine regularly
   C. all children living in Melbourne
   D. everyone living in Australia

2. Which one of the following would be the best sample for the study described in Question 1?
   A. 40 female students from various year levels attending the local high school
   B. five university students sitting on the tram
   C. 100 adults over 18 years old who consume caffeine regularly
   D. 25 children who have trouble concentrating

3. One advantage of stratified sampling over random sampling is that
   A. the sample is more representative of the population, when there is a diverse population.
   B. the sample is less biased.
   C. stratified sampling is much quicker and easier to complete.
   D. stratified sampling uses fewer participants and there is no order effect.

4. An example of a self-report method is
   A. a questionnaire.
   B. an interview.
   C. a rating scale.
   D. all of the above.

5. Which of the following is an example of a cross-sectional study?
   A. a survey of members of the Western Bulldogs Football Club before and after the grand final
   B. a survey of the number of students vaccinated for polio in Victoria in 2022
   C. a survey of the number of graduates with full-time employment in January 2020, 2021 and 2022
   D. a weekly survey of hours spent watching TV, given to 100 Melbourne residents for one year

6. Debriefing
   A. must occur after every experiment.
   B. occurs only after an experiment in which deception has occurred.
   C. does not have to occur if the researcher runs out of time.
   D. occurs only when an experiment causes distress to a participant.

7. Deception
   A. cannot ever occur in a psychological experiment.
   B. can occur only in an experiment using participants who are over the age of 18.
   C. can occur only if psychological distress is monitored.
   D. can only occur if the participants are debriefed after the experiment.

8. What must occur prior to an experiment being conducted?
   A. All participants must be debriefed.
   B. The participants must be informed about their rights.
   C. The participants must be tested for susceptibility of distress.
   D. The participants must be offered payment for their contribution to the study.

9. Which of the following best defines the term 'hypothesis'?
   A. a thoughtful, testable prediction about the likely results of a research study
   B. a prediction about the results of a research study that may or may not be able to be tested
   C. a way of collecting data on the social and emotional wellbeing of people
   D. a way of describing the results of a study, so that the conclusion is inaccurate

10 Professor Green conducted a study investigating languages spoken throughout Melbourne. He collected data from groups with different ethnic backgrounds, based on the relative percentage of the groups in the Melbourne population. This is an example of

A stratified sampling.
B random sampling.
C biased sampling.
D unbiased sampling.

11 Felicia wanted to investigate whether chimpanzees use tools to catch and collect food in the wild. She set up camp in the forest and followed a group of chimpanzees for 6 months. She found that chimpanzees use many tools in day-to-day life – much like humans. What type of study did Felicia conduct?

A case study
B survey
C fieldwork
D correlational study

12 When using informed consent, a researcher must include

A a description of the descriptive statistics that will be used.
B a description of previous research that other researchers have completed.
C the role of the participant and the hypothesis.
D the role of the participant and the rights they have while participating.

13 Which of the following are examples of quantitative data?

A height, foot size, age, temperature
B tolerance to alcohol, opinions on the prime minister, mathematical performance
C gender, city of birth, number of pets, experiences when using public transport
D IQ score, attitude to illegal drugs, VCE scores, enrolments in Victorian government schools

14 When studying the effect of a new anti-anxiety drug, participants were asked to take a pill once a day for two weeks. They were then given another bottle of pills to take for the next two weeks. In one of these weeks, the pill was a placebo. This is an example of which research design?

A independent-groups
B matched-participants
C repeated-measures
D repeated-participants

15 In the study described in Question 14, each participant did not know which week they were receiving the placebo, but the researcher did. This is an example of

A a single-blind procedure.
B a double-blind procedure.
C deception.
D counterbalancing.

Short answer

1 What is the main goal of the independent-groups design?   1 mark

2 Distinguish between an internally valid investigation and an externally valid investigation.   2 marks

3 Explain how a repeated-measures experimental design can be used to minimise the effect of extraneous variables associated with individual participant differences.   2 marks